HYPERTEXT AND THE FEMALE IMAGINARY

ELECTRONIC MEDIATIONS

Katherine Hayles, Mark Poster, and Samuel Weber, Series Editors

(continued on page 153)

HYPERTEXT AND THE FEMALE IMAGINARY

JAISHREE K. ODIN

Electronic Mediations 31

University of Minnesota Press
Minneapolis
London

Portions of chapters 2 and 3 were published as "The Edge of Difference: Negotiations between the Postcolonial and the Hypertextual," *Modern Fiction Studies* 43, no. 3 (Fall 1997): 598–630; copyright 1997 The Johns Hopkins University Press. Portions of chapter 3 were published as "Embodiment and Narrative Performance" in *Women, Art, and Technology,* ed. Judy Malloy (Cambridge: The MIT Press, 2003), 452–65. Portions of chapter 4 were published as "The Database, the Interface, and the Hypertext: A Reading of Strickland's *V,*" *Electronic Book Review* (October 2007) and as "Image and Text in Hypermedia Literature: *The Ballad of Sand and Harry Soot,*" *Iowa Review Web* (September 2002). Portions of chapter 5 were previously published as "Unraveling the Tapestry of *Califia,*" *Electronic Book Review* (September 2001) and as "Collage Aesthetics, Hypertext, and Postcolonial Perspectives," *Genre* XLI (Fall/Winter 2008): 83–104; copyright 2008 by the University of Oklahoma.

Published by the University of Minnesota Press
111 Third Avenue South, Suite 290
Minneapolis, MN 55401-2520
http://www.upress.umn.edu

Library of Congress Cataloging-in-Publication Data

Odin, Jaishree Kak.
    Hypertext and the female imaginary / Jaishree K. Odin.
        p.   cm. — (Electronic mediations ; 31)
    Includes bibliographical references and index.
    ISBN 978-0-8166-6669-0 (hc : alk. paper) — ISBN 978-0-8166-6670-6
(pbk. : alk. paper)
        1. Hypertext literature—Women authors—History and criticism.
2. Literature and the Internet.   3. Literature and technology.   4. Mass
media and culture.   5. Gender identity in literature.   6. Cultural pluralism
in literature.   7. Experience in literature.   I. Title.
    PN56.I64O35 2010
    302.230973—dc22

                                                                        2010016952

Printed in the United States of America on acid-free paper

The University of Minnesota is an equal-opportunity educator and employer.

16 15 14 13 12 11 10   10 9 8 7 6 5 4 3 2 1

The story never stops beginning or ending. It appears headless and bottomless for it is built on differences. Its (in)finitude subverts every notion of completeness and its frame remains a non-totalizable one. The differences it brings about are differences not only in structure, in the play of structures and of surfaces, but also in timbre and in silence. We—you and me, she and he, we and they—we differ . . . in the choice and mixing of utterances, the ethos, the tones, the paces, the cuts, the pauses. The story circulates like a gift; an empty gift which anybody can lay claim to by filling it to taste, yet can never truly possess. A gift built on multiplicity. One that stays inexhaustible within its own limits. Its departures and arrivals. Its quietness.

TRINH T. MINH-HA, *WOMAN, NATIVE, OTHER*

CONTENTS

PREFACE

Technological shifts and cultural shifts mutually shape each other while at the same time promoting new experiences of subjectivity and agency that shape and are shaped by modern narratives. In *Hypertext and the Female Imaginary* I explore representative works by postmodern women writers or artists who address issues related to gender or cultural identity in local and global contexts while bringing out the significance of difference in an increasingly homogenized and homogenizing global culture. I frame my discussion in the broader technological shift that has given rise to new constructions of subjectivity and agency. In information culture, biological and social metaphors are used to describe intelligent machines, or humans are equated with computers. As metaphors move back and forth between scientific and cultural narratives, they naturalize disembodied constructions of the human. The construction of the "disembodied" human in the technological and cultural narratives makes the need (or, for that matter, the desire) for a politics of social change seem superfluous.

The underlying assumption of this study is that a paradigm shift is required in order to bring attention to the embodied status of the human and the situated nature of experience. The shift involves seeing the contemporary field of experience as a topological space where different worlds intersect at different levels, requiring new modes of representation to convey the complexity of postmodern experience. The perpetual negotiation of difference across worlds and across cultures produces its own dynamics, which is best conveyed through hypertext or collage techniques of discontinuity, fragmentation, multiplicity, and heterogeneity. The electronic media easily lend themselves to creating complex narratives of multiple worlds or worlds within worlds that have potential for diverse trajectories of meaning. To what the media make possible with respect to accessing the text is added another dimension, the role the media now play in the meaning-making process.

I thank my colleagues Peter Manicas and Emanuel Drechsel, who have supported my research and teaching over the years in the area of cultural studies of science and technology. My journey in this field began about two decades ago when I came across N. Katherine Hayles's thought-provoking books *The Cosmic Web* and *Chaos and Order*, which explored the interdisciplinary relations among literature, science, and culture. I became interested for the first time in joining two aspects of my training, my graduate work in chemistry with my postgraduate work in comparative literature. Soon afterward I had the opportunity to attend Hayles's 1995 National Endowment for the Humanities seminar, "Literature in Transition: The Impact of Information Technology" at UCLA, which introduced me to hypertext literature and the work of hypertext pioneers Michael Joyce, George Landow, Jay Bolter, Judy Malloy, Shelley Jackson, and Stuart Moulthorpe, among others. Many of the seminar participants, including Joseph Tabbi, Paul Harris, Stephanie Strickland, and Marjorie Luesebrink, continued to work in this area over the years, either producing hypertext works or writing about electronic literature. Their work and that of many other hypertext writers and artists with whom I have interacted over the years at conferences has been a constant source of inspiration and new insights as I worked on this book.

INTRODUCTION

Contact Zone: Repetition and Difference

NEW TECHNOLOGIES HAVE PROFOUNDLY IMPACTED literary as well as artistic practices as old media partially or fully adopt the strategies and conventions of the new media and vice versa. A fundamental transformation has also occurred in artistic perception regarding how an artistic work is to be conceived and created and what function it serves in broader society. In a complex shifting world, artists and writers are no longer in a position to convey the truth about human existence or the world they inhabit. The current generation of writers of critically aware fictions oscillates between the need to explore the tools of their trade to get readers actively involved in the fiction-making process and the necessity of keeping their fictions grounded in the social reality; the need to expose the constructed nature of the fictional world and the necessity of these structures; and the need for belief in truth, knowledge, and order in the world and the necessity of resistance to any predetermined definitions of these notions. Theorists have called the shift in literary representations of the last few decades radical realism, which combines realistic details while skillfully incorporating postmodern narrative strategies to depict experience that is profoundly affected by the scientific and technological reorganization of sociopolitical, cultural, and economic conditions of production and consumption.

In *Hypertext and the Female Imaginary* I explore representative postmodern works, in primarily electronic and filmic media, that use hypertextual strategies of narrative fragmentation to engage questions of gender or cultural difference. I use the term *hypertext* specifically for electronic or film narratives in which discontinuity is a major artistic strategy, but I also use it as a metaphor for describing the complexity of postmodern culture, where different cultures, discourses, and media are in constant interaction with one another against the background of technocapitalism. My readings are from four main perspectives—discontinuity, fragmentation, multiplicity, and assemblage—that characterize postmodern topology. Even though all these perspectives appear together in critically aware works with fragmented narrative surfaces, I explore each perspective individually with respect to a specific postmodern work, whether in film or digital media. A thread that connects the individual chapters is the role of hypertextual strategies in opening up the narrative space for other stories and other voices to surface. I primarily focus on the mutation of narrative as seen in works created by women, for example, Trinh T. Minh-ha, Judy Malloy, Shelley Jackson, Stephanie Strickland, and M. D. Coverley. However, I also attempt to go beyond the mutation of narrative to see if such narratives can serve as a counterdiscourse to the homogenized and homogenizing dominant technocratic narrative through bringing out the significance of difference, whether cultural or gender, as it is enacted rather than as it is represented.

An important question that I explore through my readings of various transformative hypertext narratives is what constitutes empowering descriptions of the world in a technology-mediated culture in which all aspects of life are increasingly shaped by technocapitalist forces that fragment communities as well as the spaces they inhabit. In a world where cybernetic constructions of humans as well as texts are pervasive, a paradigm shift is required in order to bring out the significance of the situated nature of experience and the interconnectedness of humans and the worlds they inhabit. The shift involves seeing the contemporary field of experience as a topological space in which different cultures and cultures of science and technology intersect at different levels and create a dynamic space that shapes and is shaped by humans.

As a counterpoint to women's narratives I provide a reading of Neal Stephenson's *Diamond Age* in which he addresses the broader questions of how to bring about positive change in a technically advanced global society in which all values have been transformed into market values. Stephenson's narrative points to the need to adopt successful social models to create an orderly society and the need to use computer simulations to teach young people the wisdom encoded in the universal archetypes of various cultural traditions. But a culture, whether old or new, cannot be reduced to a database to be made available to youngsters on interactive devices, and historical social models are not robes that individuals or groups can wear to realize a particular future. The cultures as they have evolved are a result of the dynamic interaction of a variety of forces historical, political, and economic as well as social, and as such they represent topological spaces of experience. Stephenson's perspective has little transformative value because it is guided by the dominant narrative based on the cybernetic construction of the human.

Such narratives, pervasive in contemporary culture, make me circle back to my central argument, that to resist the dominant techno-rationalist configurations in society as well as cybernetic constructions of the human it is essential to move away from theories that reduce humans to intelligent machines or to genetic information and reduce cultures to databases that can be transmitted to future generations on interactive devices. In order for genuine change to occur, the situated nature of experience and the embodied status of the human must be emphasized. Such a perspective brings out the significance of other stories and other perspectives as they are enacted rather than packaged and put into the circuit of production and consumption. Otherwise, the so-called transformative narratives simply remain as new wine in old bottles.

Contact Zone and Difference

In contemporary narratives representation as the basis of narrative-making has given way to the actualization of the narrative occasion that comes into being in the act of reading. The creation of performative narratives in both electronic and print media has been directly linked to the reconceptualization of the notion of difference in theoretical writings as well as cultural practices of the last few decades. In *Difference and Repetition* Gilles Deleuze (1994) argues that the notion of difference arose in the west within the context of representation, which has been the basis of western systems of knowledge in which the starting point of reference has always been the absolute, God, or Plato's ideal forms. Such dualistic frameworks operate in a system of identity and difference in which the secondary term or text is formulated in terms of its similarity or difference to the first term or text. Deleuze rejects any transcendental unifying principle or ground of experience by denying "the primacy of original over copy, of model over image" (66). When there is no original or copy, the notion of representation based on a system of identity and difference loses its validity.

Deleuze's philosophy privileges difference in order to provide a theoretical framework for understanding the postmodern condition. Moving away from Leibniz's notion of the individual as a "monad" reflecting the perfect world of God, though only imperfectly, Deleuze turns it around to serve his own purpose. In his theoretical framework the monad is transformed into the "nomad," who is always in the state of changing into something else. In nomadology the telescopic multiple perspectives of the one world in the universe are thus replaced by the multiple perspectives of multiple worlds that actualize as innumerable diverging and converging series. The individual as nomad simultaneously partakes of multiple worlds and worlds within worlds. Drawing on the expressionist philosophy of Spinoza, Deleuze describes the concept of difference as it functions in terms of expression. The expression cannot be separated from what is expressed. Similarly, in a chaotic and fragmented world, signs no longer point toward a reality "out there," but rather stand for the object that they create and erase at the same time. This has important implications for critical practice as the attention shifts away from a search for meaning in life or literature to see how the nonorganic force we call life actualizes itself in individual life as well as literature.

Using theater as a metaphor, Deleuze elaborates on how contemporary knowledge can be seen as based on a system of repetition rather than representation. In a theatrical performance the actor does not move from a particular concept to its representation in performance; rather the actor performs the role and within the theatrical space becomes the role. As a new play is performed, the actor moves to a different role that might or might not have anything to do with the previous role. The relationship between the roles performed is not that of opposition or mediation but simply that of contiguity in which the act of performance is a repetition and difference rather than identity and difference. The actualization or the repetition of the invisible and the indeterminate in the visible illustrates the epistemic shift from dividing reality into two opposing states that are ontologically different to their coalescing into a single state. One could say that "the bottom" or the depth dimension is repeating or actualizing in the surface rather than represented in the surface.

Deleuze uses Borges's classic story "The Garden of Forking Paths" (1962) to make clear the epistemic shift to the performative framework. In the story Feng's ancestor Ts'ui Pên from China leaves a legacy behind in the form of his labyrinthine project called "The Garden of Forking Paths." Albert, the scholar, realizes that the project refers to the chaotic novel that Ts'ui Pên had spent years writing, which represents his conception of the universe. In a traditional narrative, if a character is faced with several alternatives, he picks one of the alternatives and the narrative proceeds along that path. But in Ts'ui Pên's fictions "he chooses—simultaneously—all of them. *He creates,* in this way, diverse futures, diverse times which themselves also proliferate and fork" (26). Ts'ui Pên is thus not interested in representing just one perspective or a single storyline—he wants to convey the experience of different storylines as they are enacted in their difference. The actual that materializes in a particular narrative trajectory does not resemble or represent the virtual possibility that it embodies. The notion of the actualization of the possibility is different from the concept of the representation of the possibility. The latter involves a conceptual mediation between the subject and the object based on identity. In a performative knowledge system, however, both subject and object retain their uniqueness and the relationship between the two is based on difference rather than identity.

Whereas Deleuze presents a general theory of difference in a world which no longer lends itself to description in terms of systems of representation, Luce Irigaray (1985) is more specific and focuses on the need for a theory of sexual difference in western systems of knowledge that are based on sexual indifference. She argues that the western representational systems, both discursive and symbolic, work on the "logic of the same," which means that all elements are seen in terms of their similarity to the first principle. When the first principle is male-defined, everything outside the first principle becomes feminine or the other, not only in discursive and symbolic representations but also at the concrete level, on which women and nature are classified as the other. According to Irigaray, the suppression of the female imaginary in language and in the symbolic order of myths, literature, and art has prevented women from experiencing themselves as speaking subjects. The sexual indifference, she adds, must be replaced by sexual difference. Instead of one sex seeing itself reflected in the other, both men and women should experience themselves as subjects reflected in the othering of their own selves rather than the other selves. Irigaray resists giving any essentialist definition of the female subjectivity by employing elemental imagery of the cave, the womb, or fluids when she speaks about the repressed feminine. The emphasis in her writings is more on the concept of inclusion of difference itself than on creating essentialist male and female categories.

Other feminist theorists have commented on the performative nature of gender. Although nobody would deny that the body is the stuff of concrete physical matter, the materiality of the body, which determines how we are oriented toward the world in both material and psychological terms is very much determined by our location. The dominant culture inscribes stereotypical gender roles, which are then incorporated into cultural practices. Judith Butler argues that the materiality of the body is a construction that emerges out of a field of power that shapes its contours, marking it with sex and gender. She points out that we need to rethink the very meaning of construction and the grammatical structures that we use when we talk about construction. For her it is "neither a single act nor a causal process initiated by a subject and culminating in a set of fixed effects. Construction not only takes place *in* time, but is itself a temporal process which operates through the reiteration of norms" (10).

To describe the materiality of the body as a construction in Butler's theorizing, then, is not to resort to linguistic determinism or cultural constructivism. We take it for granted, Butler notes, that somebody—that is, a human subject or, in more recent formulations, something such as culture, discourse, or power—does the act of constructing. In the first we resort to metaphysical claims, assuming that there is a subject that exists prior to any sociocultural induction, and in the second we forfeit the agency of the subject and replace it with a surrogate agent in the form of culture, discourse, or power. Rejecting both claims, Butler describes the materiality of the body as arising in a matrix of power relations so that the agency of the subject comes after, not prior to, the materiality of the body, emerging through a process of enactment. "To claim that the subject is itself produced in and as a gendered matrix of relations," Butler notes, "is not to do away with the subject, but only to ask after the conditions of its emergence and operation" (7). By reformulating the very meaning of construction, locating it in time, and describing it in terms of a temporal process, Butler reveals the constructed nature of naturalized states of sex and gender. The performance of gender is thus a constant reiteration of the regulatory norms, and it is in the performance of these norms that the materiality of the body emerges. To unearth the female imaginary to serve as the counter-discourse to the patriarchal regulatory norms is thus part of what women artists have felt they need to do in order to fully realize their artistic potential. Judy Malloy's *its name was Penelope* (1993) and Shelley Jackson's *Patchwork Girl* (1995), discussed in a later chapter, are about unearthing women's literary history, or the lack thereof, so the present can be experienced with and in difference.

Although women writers and theorists bring out the significance of gender difference and the situated nature of experience, post-colonial theorists have commented extensively on cultural difference and its performative nature. Minority cultures continuously negotiate with the dominant culture, which is increasingly defined by technocapitalist configurations that empower some and disempower a vast majority. Some theorists have named the space of encounter in which minority cultures meet the dominant culture the "contact zone" or "third space of encounter" (Pratt 1992; Bhabha 1994). *Contact zone* describes the experience of in-between spaces, which are states of becoming that cannot be categorized or fixed into any stable final formulation. At the dominant site the performance involves the repetition of the same, whereas in the contact zone the repetition takes place with and in difference. The "contact" subject, then, is the processual subject who must at each moment negotiate difference and whose essence lies in the act of becoming rather than being. The performance of the same with and in difference challenges naturalized dominant social and cultural norms.

Mary Louis Pratt and Homi Bhabha use the terms "contact perspective" or "third space" in the context of formulating a postcolonial response to the experience of colonialism, but these notions can be equally applied to the global colonization of minority cultures by the dominant technocratic narrative. Different cultural traditions are not fixed in time, and any articulation from within these traditions must involve a complex and ongoing negotiation with spaces marked with disjunctures and contingencies that resist reification into categories. Bhabha notes that the repetitious and recursive nature of the performative appears as cultural displacement in which repetition at different sites is accompanied with difference so that any homogenizing descriptions are impossible. A meaningful experience of cultural history involves a reenactment of the signs and symbols in the form of repetition or reenactment in which signs are experienced directly, without any mediation, connecting the subject to the history, language, and symbols of the culture. The movement from the past to the present or from the symbols, myths, and stories to their reenactment is not that of the representation of some original story, absolute truth, or essence, but the repetition itself constitutes the fabric of the continuity of the symbolic order of a culture.

In an increasingly commercialized world, cultural diversity is often turned into a commodity to be circulated for consumption.[1] The notion of "cultural difference" is vastly different from the generally held conceptions of multiculturalism and cultural diversity. Whereas both "multiculturalism" and "cultural diversity" are based on cultural relativism, the notion of cultural difference is grounded in a performative theoretical framework that emphasizes the role the cultural or community history plays in the daily lives of people, not only at the discursive level but also in the lived reality manifested in the bodily practices of the people.[2] For this reason, Trinh T. Minh-ha, a writer and filmmaker, rejects the notion of "authentic culture" as perpetuated by ethnographic films. In her films on West Africa, *Reassemblage* (1982) and *Naked Spaces—Living Is Round* (1985), she focuses on bringing out the complexity of the daily lives of the villagers from multiple perspectives without framing these perspectives in the dominant western perspective.[3] Trinh uses hypertextual strategies to present a montage of cinematic images and sounds in such a way that the viewers themselves are invited to get actively involved in the meaning-making process. Instead of presenting the villagers' experiences as packaged images for consumption, the camera records people as they go around performing their daily activities. Trinh's films convey the topological spaces of experience as she captures the lives of villagers in the daily act of living without providing any commentary.

Topographical Maps, Topological Spaces, and Narrative Mutation

At this point it is useful to step back and contextualize the postmodern experimentation with narrative fragmentation in the broader literary history in order to understand the shift from representation to performance or from the mirror metaphor to that of the silverless mirror. The perception of literature and art as mirrors of reality was shaped by advances in science and technology. As the science of optics developed in the seventeenth and the eighteenth centuries, the exact representation of the real world became the basic tenet of scientific thinking. The invention of the telescope and the microscope allowed the very small and the very distant to be grasped through the human eye. In literature this worldview led to the development of the realistic novel with an omniscient narrator who, through an all-seeing eye, had knowledge of everything happening in the narrative.

Ocular metaphors and mirror imagery also constituted an integral part of traditional western philosophy. Descartes's question "How do I know that the representations of things I see correspond to the things out there?" led to a line of questioning that turned sense faculties into conduits of information that connected the external world to the internal world. The human subject became located somewhere in the brain, viewing the representations of the world reflected in the mirror of the mind. The Cartesian philosophy personified the dualistic worldview that was to deeply influence western systems of thought for centuries to come. The eye was thus severed not only from the body but also from the world in which it was located. The ocular metaphors, Richard Rorty writes, influenced western philosophy's preoccupation with epistemology instead of ontology and also led to the increasing separation of the mind from the body.

With the advent of the twentieth century, there was a shift in thinking as developments in psychology, linguistics, physics, and mathematics pointed toward a multilayeredness of existence that could not be reduced to any single perspective. In psychology Freud's theories explored irrational desires and fears shaping human actions, and in linguistics Saussure's theories postulated the functioning of language as a system of signs whereby the meaning of each sign is determined by its relationship to the other signs in the system. Gödel's theorem in mathematics showed the self-referentiality of mathematical language. In physics Einstein's relativity theory resulted in the replacement of the static and mechanical model of the universe based on the Newtonian worldview with a new model that described the universe as an interacting system in which space was not a container for the progression of time or events but rather timespace constituted the fourth dimension of reality. The development of quantum mechanics added the element of uncertainty to all descriptions of reality at the subatomic level. Niels Bohr and Werner Heisenberg were pioneers of metaphysical speculations that emphasized the indeterminate nature of knowledge at the subatomic level. Based on the developments in physics and mathematics, a field model of reality emerged that was different from the earlier Newtonian model. The Newtonian worldview based on division and analysis makes it easy to take for granted the objective position of the observer to describe a system accurately using neutral language. The field model of reality, on the other hand, recognizes the interconnected nature of the parts constituting a particular system that includes the subject, the object, and the language used to describe the relationship between the two.[4]

In the late nineteenth and early twentieth centuries modernism in literature and the arts was an exploration on two fronts, first to develop strategies for representing the multidimensionality of experience in an interconnected world and secondly to test the limits of language, visual or verbal, to convey a complex subjective reality in its immediacy in order to arrive at the essence of human experience. Thus writers took to representing the subterranean depths of consciousness, most noticeably in the psychological and stream-of-consciousness narratives of the first half of the twentieth century. The fragmented discontinuous representation of reality replaced the linear causal narratives of realist fiction. But behind this discontinuity, fragmentation, and chaos in the world was the modernist belief in a unified and complex underlying reality. The writers and artists turned their gaze inward to capture the eternal and the everlasting in a world where everything seemed ephemeral. The inner eye that artists and writers used to convey their vision displaced the outer eye valued in realist fictions. Their fictions explored the expression of knowledge as a universal category, along with the process of communicating and representing it from a variety of perspectives. In the process, modernist writers became preoccupied with language and their fiction showed increasing self-referentiality. A modernist work no longer functioned as mimetic representation of the real life or as the artist's self–expression; it became an artifice with its own reality.[5] James Joyce, Marcel Proust, and Virginia Woolf, among others, tried to capture the spirit of the times in their works.

It is useful to note here that even though modernist artists and writers focused on subjective experience and turned away from mimetic representation of the earlier realist fiction, there was continuity between the two movements in that the underlying assumption from which both operated was their belief in a universally valid metaphysical basis for knowledge and experience. That certitude was lost in the 1960s with the emergence of counterdiscourses of feminism, poststructuralism, and other minority movements that showed that all grand narratives—historical, philosophical, psychological, and even scientific—were stamped by the historical location of their creators. As the metaphysical foundation on which the humanist as well as scientific discourses rested gave way, so did the belief in the grand narratives of progress, equality, and emancipation. Lyotard terms the shift in the construction of knowledge a shift from grand narratives to local narratives. The shift had an immense impact on the development of literature and created a crisis in representation. In the 1960s and 1970s there was intense experimentation with the language itself, resulting in self-reflexive fiction that lost itself in its own coming-into-being. Writers sought to find their own unique ways to discover alternatives to mimetic representation as they explored formal techniques, leaving behind mimetic conventions of representation.[6]

Early postmodern writers thus focused on formal techniques and produced self-reflexive fiction as their fictions turned inward to the linguistic labyrinth with no center or foundation, no origin or end. In spinning out fictions about writing fiction, they smashed the mirror of representation into pieces, and out of these emerged new self-reflexive worlds that expressed the crisis of a generation of writers. For experimental writers self-reflexiveness was an expression of their disbelief in metanarratives of all kinds. John Barth, Donald Barthelme, Robert Coover, Walter Abish, Italo Calvino, Julio Cortázar, and Jorge Luis Borges show a fascination with writing about writing that manifests itself either in terms of self-reflexiveness or in terms of fictions that strip their connection with reality. Writers of this era expose the constructed nature of the artifact while at the same time exposing the constructed nature of social, political, moral, and aesthetic values.

Italo Calvino's *Cosmicomics* (1968) and *t zero* (1969) collection of stories create fantastic fictions about the birth of the universe weaving fiction out of the scientific hypothesis, thus pointing to the relative nature of both types of narratives. Another example of the self-reflexiveness that liberates fiction from illusionism and fictionality is Paul Auster's *City of Glass* (1987), in which the world appears fragmented and disjointed and the language no longer serves the function of connecting people with one another or with the community. The turning in on itself becomes a narrative strategy to problematize the effectiveness of language either to represent the material reality (realist preoccupation) or to express the inner reality (modernist preoccupation). The writers of this era thus show fascination with multiple worlds, worlds within worlds, interweaving worlds, colliding worlds, and textual labyrinths.[7]

Calvino's philosophical prose poem *Invisible Cities* (1974) is perhaps the best example to represent the paradigm shift involved in moving from a modernist to a postmodern perspective or from a topographical to a topological perspective in narrative construction. Calvino's text has been described as a rewriting of *Le travail* of Marco Polo, the thirteenth-century Venetian who traveled in Asia for many years and even served as an ambassador to the court of Kublai Khan. In Calvino's retelling, the Khan sends Polo to visit the cities in his empire to bring back stories about the land that he has conquered. Polo's stories fascinate the Khan because they are different from what he hears from his other messengers, who stick to factual details about the land, taxes, and the numbers of the dead and living in the Khan's growing empire. Polo does not give topographical accounts of material cities grasped by the eye. His tales reflect and refract relationships between each city and its inhabitants and between the city and the traveler passing through it. Polo captures the forces, desires, and fears that move cities forward in time, make them grow, reach their pinnacle, and decay.

As Kublai Khan listens to Polo's account of the cities, he starts creating imaginary cities in his mind. At times the Khan thinks that he is on the verge of discovering a coherent, harmonious system underlying the infinite deformities and discords in his empire, but none could stand up to a comparison with the game of chess. As he comes to this conclusion, he sees no need to send Polo on distant travels.

All he needs to do is to play a game of chess with him to find out more about his empire. He uses the movement of the pieces and the ups and downs of the game to understand the condition of the cities. As he plays game after game, the purpose of the game eludes him. "Each game ends in a gain or a loss: but of what? What were the true stakes? At checkmate, beneath the foot of the king, knocked aside by the winner's hand, a black or a white square remains" (123). The Khan thus disembodies his conquests and reduces them to their essential states; the treasures of his empire appear to him as an "illusory envelope," like the square of wood on the chessboard, which ultimately means nothing. Polo once again tries to correct the Khan's perception by bringing to his attention the history of the chessboard. Whereas the Khan dissolves the chessboard and the games played out on the board into nothing, Polo takes the opposite view and reveals the materiality of the chessboard and the stories it contains of its own genesis as an artifact. The material history of the chessboard brings together images of the ebony forests, the women at the windows above the harbor, and the rafts laden with logs going downstream.

Calvino's *Invisible Cities* creates a performative space that the reader can enter and explore in multiple ways. This vision is also reflected in the construction of the text, which can be read in any order—in a linear fashion from beginning to end, starting from the end and proceeding in the opposite direction, or even starting from the middle. The narrative structure is elaborately planned. Each of the nine chapters contains an opening and a closing dialogue between Marco Polo and Kublai Khan. The nine framing dialogues turn the group of cities in each chapter into a framed collage of multiple cities. The first and the ninth chapters include ten cities each, whereas the intervening seven chapters contain short descriptions of five cities each. The cities are divided into eleven categories, for example, cities and memory, cities and desires, cities and signs, trading cities, cities and names, cities and the dead, and so on. The total number of cities adds up to fifty-five, and with the addition of nine dialogues that accompany each chapter, the number jumps to sixty-four, which is the same as the total number of squares on a chessboard. The seemingly chessboard architecture of the work is misleading, however, because the exchange between Polo and the Khan reveals that cities have a logic of their own that is ingrained in their histories, their perspectives, and the desires and fears of their inhabitants. This individual perspective embodied by each city cannot be reduced to a set of abstract rules by which to deduce its current or future states of existence, its birth, decay, or rebirth.

Polo's tales of the cities reveal that he reaches out into the past and moves forward into the future at the same time, thereby enriching his present. The Khan, however, is a figure from the past who stays stuck in the present, which is disconnected from both the past and the future. The two figures confront each other, and a third narrative emerges as the reader's view is opened up to the plurality of the text. If the dialogue between the Khan and Polo illuminates anyone, it is not the Khan but the perceptive reader, who soon realizes the tension between the two juxtaposed discourses that constitute the invisible center of the text. Polo's comment "It is not the voice that commands the story: it is the ear" (135) is meaningful in this context because it reflects on the text itself in that the Khan's understanding of the tales is different from that of a contemporary reader as the latter has the hindsight of centuries to shape the reading.

In exploring how *Invisible Cities* functions, the reader discovers that any city, whether of text or of material reality, can be approached from a multiplicity of perspectives because it constitutes a topological space that includes the material reality of the city, the inhabitants who occupy it and the socioeconomic and political forces that shape it. The collage principle that permeates the text ensures that no single meaning can be reached behind the multiplicity of cities. The function of the reader is thus to unravel the relationships between different categories of cities and, within and outside each category, the relationships between the cities themselves and finally between the cities as they relate to the frame dialogue.

In *Invisible Cities* Calvino uses narrative discontinuity to reflect on the paradigm shift from modernism to postmodernism. However, minority writers such as Salman Rushdie, Toni Morrison, and Leslie Silko, among others, have used discontinuity as a major artistic strategy to create print narratives with a political edge. In the historical metafiction of Salman Rushdie, Linda Hutcheon (1989) writes, the postmodern reappropriation of historical materials is done to insert the past in the present so the present can be reseen and reevaluated with and against the past, which in some cases has been forgotten and in other cases suppressed. The historical continuity is punctuated by discontinuities, and it is along these fractured lines that histories of others are revealed. Thus "story telling is not presented as a privatized form of experience but as assessing a communication bond between the teller and the told within a context that is historical, social and political as well as intertextual" (50–51).

Writers in the last few decades have increasingly turned to these fissures and fractures to give shape to their personal histories as well as community histories even as there has been an awareness that history as well as the perception of history changes as society changes. Leslie Silko (1996), for example, sees the urgency of reviving the stories of her people and their actual encounters with these stories in order to enable identity as well as politics. Her stories are grounded in the Pueblo Indian oral tradition, which encodes the cultural experience of her people. She notes that "the ancient Pueblo people depended upon collective memory through successive generations to maintain and transmit an entire culture, a worldview complete with proven strategies for survival. The oral narrative, or story, became the medium through which the complex of Pueblo knowledge and belief was maintained" (30).

Silko translates the oral tradition into her written narratives, retaining the ambiguity, participatory nature, and multiple, sometimes contradictory, meanings of oral narratives. In *Ceremony* (1977), she hypertextually punctuates the contemporary narrative of the protagonist's experiences with a folktale throughout the text. The fragments of the purification ritual encoded in the oral tale correspond to different stages in the protagonist Tayo's healing, so that the ritual ends around the same place at which Tayo's healing is complete. The two narrative threads can be read separately, or the oral narrative can be seen as a thread woven through the written narrative. The textual play of two narratives—the old and the new, the past and the present, the oral and the written—creates a performative space whose reality emerges in the act of reading.

Although *Ceremony* is tightly structured, Silko's *Storyteller* (1981), in contrast, can be seen as a loosely spun web of the old Laguna tales that she heard as a child. The Yellow Woman stories appear and reappear quite frequently in this text. The various versions of the old Laguna tales describe the fate of the Yellow Woman differently. The blending of the mythical tales with the events in the narratives makes the present events a reliving of the old tales but with a difference that is shaped by the location of the narratives—in both the historical and the sociopolitical context. The telling and retelling of the tales continually blend them in the day-to-day narratives of the people, involving them in a complex ongoing negotiation with the dominant culture. The individual stories in *Storyteller* resemble interlinked hypertextual units or lexias that are interrelated, each story linked to every other story in the text. Silko's fragmentation of the textual surface by incorporating the oral tradition into her narrative allows her to effectively convey the performative nature of both minority experience and subjectivity.

Toni Morrison also uses strategies of discontinuity and fragmentation in *Beloved* (1988) to create a fragmented narrative space that brings together different worlds that on contact with one another coalesce into novel configurations. The text is permeated with cinematic or visual images as well as techniques used in black music.[8] The visual cuts continually break the narrative, resulting in juxtaposed images, sometimes mirroring or inverting the same themes, and sometimes bringing together opposing images. Thus *Beloved* "chronicles not a list of incidents, but a web of interconnected moments. Events and moments of awareness frequently emerge in mirroring forms, in inversions, parodies and repetitions. The separate scenes of the text are woven together in a montage" (Koolish, 424). The montage of images and scenes creates gaps in the narrative that are an invitation to the reader to actively participate in the telling and untelling of the stories. *Beloved* is thus not a traditional slave narrative but rather an assemblage of points of view through which the experience of slavery and what came before it is reflected as well as refracted and its place in the African American constitution of subjectivity is explored.

In *Beloved* Morrison uses memory as a device with which to access the social and psychological forces that constitute the subjectivity of black people in America from both historical and contemporary perspectives.[9] The privileged moments are recollected and retold from multiple perspectives, which sometimes turn out to be variations of the same event. For example, the story of Denver's birth is told and retold several times; each recounting is a variation of the one told before. The repetition in *Beloved* reflects that in the oral tradition it is the ability of one story to multiply into multiple stories that reflects a culture's ability to rejuvenate from within. The characters capable of change and growth are finally those who can remember or narrate multiple versions of the same story rather than those who turn all stories into one story. This flexibility becomes a tool for resisting the dominant discourse and allowing people to become the subjects of their own stories.

Morrison's unique style incorporates the black oral tradition in a way that brings about a displacement between the physical and the psychological, life and death, the African and the African American past, the world of objects and the world of shadows, the real events and the memories of events, factual accounts and fictive accounts, and even the terms *black* and *white*.[10] *Beloved* is finally an expression of the complex interrelations of African Americans with the present and the past whereby the unknown and the unfamiliar, the forgotten and the repressed, are brought to the surface so learning and apprenticeship can take place. Morrison expects her readers to take responsibility for making sense of *Beloved* and how the narrated past as well as the expressive present, forgotten and yet remembered, always present yet absent, is to be integrated into the present for African Americans so that they can reconstitute their subjectivity.[11] Morrison's writing is thus not simply the expression of individual experiences of a writer; it invariably draws in the larger circles of the community, the village, the broader world, and the driving forces behind these worlds. It is political in nature because another story— that of the group or the community—vibrates within the stories of the individual.[12] Reading such works becomes an ethical task as readers are given the responsibility for deciding what they want to make of the recuperated past.

Because the minority experience unfolds at the pressure points of the social and the political, critically aware minority writers use discontinuity and fragmentation to create a performative narrative space with gaps to give voice to stories that are suppressed by the dominant narrative. In spite of experimentation with nonlinearity and discontinuity in print narratives, this space is limited by the physical constraints of the medium. In electronic hypertext narratives writers have explored nonlinearity and discontinuity to the greatest extent because of the flexibility of the electronic media and the mixing of the media, which can result in fascinating multimedia texts that bring out the multidimensionality of a culture from a variety of perspectives.

Nonlinearity, Discontinuity, and Hypertext Literature

The first-generation literary hypertexts created in the late 1980s and early 1990s represent the euphoria of working in a new medium that provided freedom from the shackles of linearity and fixity of the print medium. Early hypertext writers broke the narrative into small chunks of verbal units that were then connected through an elaborate system of linkages. The planning of links required complex cognitive mapping to create pathways for the reader. The reader accessed the narrative segments one at a time by clicking on the links. Many hypertext writers provided detailed overviews or maps of their narratives to guide readers. Some of the early notable hypertexts are Michael Joyce's *afternoon, a story* (1987), Stuart Moulthrop's *Victory Garden* (1992), Carolyn Guyer's *Quibbling* (1992), Judy Malloy's *its name was Penelope* (1993), and Shelley Jackson's *Patchwork Girl* (1995).

Michael Joyce's classic hypertext *afternoon, a story* (1987), in the hypertext editing program Storyspace, consists of 539 lexia, or textual units, and 950 links. In an insightful evaluation of the text, Bolter (2001) writes that "*afternoon* is about the problem of its own reading.

In its attempt to redefine the act of reading, *afternoon* has served as a paradigm for much of the early hypertext fiction" (128). In order to access the text, the reader clicks on the links encoded in the individual words of the narrative units. The choices that the reader makes determine the narrative trajectory that materializes on the screen. As the reader goes through the text, clicking on the links, a narrative emerges about Peter's search for his former wife and son, who have apparently been involved in an accident. The text has many recursive loops that prevent the reader from arriving at a definitive reading in a single session. Joyce's story has a detective-story flavor in that the reader encounters the screens of disjointed narrative text that must be put together in one's imagination to arrive at the most logical sequence of events. Joyce's *afternoon* is a narrative without a closure, because there are multiple narrative possibilities, sometimes contradictory in nature, inherent in the text.[13] The force that propels the reader to continue reading is the desire to find answers to the questions that the narrative poses. This is probably the reason that some critics have compared *afternoon* to an adventure game and a cybertext (Bolter 2001; Aarseth).

Electronic literature has evolved over the last two decades as more software applications with multiple functionalities have become available. Some new-media writers and artists focus on the verbal text as the centerpiece, with the links constituting an important aspect of accessing the database. Others use the programming potential of computers to create text generators (poetry generators, automatic poetry, and so on) whereby programmed elements determine what appears onscreen and in what sequence. Yet others create elaborate navigational interfaces whereby both the verbal text and visual displays become an integral part of the reading experience. Caitlin Fisher's *These Waves of Girls* (2001), an associative hypertext, is a web of memories created through chunks of text linked with images. Talan Memmott's *Lexia to Perplexia* (2001) approaches the question of web-based writing from a totally different perspective as he meditates on the coming into being of words and sentences as codework that reflects the coalescence of theory and fiction. M. D. Coverley uses hypertextual strategies in her hypermedia works *Califia* (2001) and *Egypt: The Book of Going Forth by Day* (2006), to create a literary assemblage through mixing text, images, and sound to create an interface that is as much a guide as a means to access the database of diverse narrative elements ranging from music, artwork, and fiction to history, myths, and legends. Her works bring out the significance of forgotten cultural narratives that can have a transforming impact on the present. Whereas cultural differences are an important part of living in a complex world, there are other perspectives that intrude into postmodern existence, including new insights gained through advances in science and technology. It is this broader perspective that Stephanie Strickland addresses in her print works "The Ballad of Sand and Harry Soot" (1999) and *V: WaveSon.nets/Losing L'una* (2002) and in her coauthored electronic works based on these print texts. Strickland's vision is expansive; she focuses on how the multiple discourses of science, mathematics, poetry, philosophy, and biography shape contemporary experience as well as artistic vision.

Several theories have been proposed for understanding the working of new media literary works to overcome the limitations of earlier hypertext theory, which focused primarily on the nonlinearity of the new form of textuality. Espen Aarseth presents his cybertext theory as a universal theory of textuality applicable to a wide range of "ergodic" texts including hypertext, textual adventure games, computer-generated narratives, participatory world simulation systems, social textual MUDs, print-based combinatorial works like the Chinese classic *I-Ching*, Raymond Queneau's sonnet machine, and *Cent Mille Milliards,* among others. He describes an ergodic text as a textual machine that comes into being through the interaction of verbal signs, material media, and a human operator. The cybernetic circuit in a textual machine constituted by the three components has boundaries that are "fluid and transgressive" because "the functional possibilities of each element combine with those of the two others to produce a large number of actual text types" (21). Because Aarseth focuses so exclusively on the functional and semiotic aspects of the narrative process, he leaves out the aesthetic dimensions of a text. His notions of "event space," "aporia," and "epiphanies" are applicable to computer games, poetry machines, or other combinatorial print works, or even to some first-generation hypertexts whose emphasis is on investigative exploration. These notions are in need of some revision to be applicable to literary hypertexts in which the aesthetic dimensions of the work leading to epiphanies of more traditional kinds are more significant than epiphanies of the reader or detective or player in search of solutions to make the work move forward.

N. Katherine Hayles's "media-specific analysis" as presented through the fictional persona of Kaye in *Writing Machines* (2002) overcomes some of the limitations of Aarseth's model by being more sensitive to the rich contextuality of the embodied text. Kaye describes all literary texts that interact with their own materiality as "technotexts" with three characteristics: chunked text, links, and multiple reading paths. Technotexts can be seen as writing machines that give physical form to both content and artistic strategies that the user sets in motion though her interaction with the interface. The materiality of a technotext is an emergent property that comes into existence through the interactions between the physical properties of a work and its artistic strategies. For this reason it cannot be specified in advance. In other words, "materiality depends on how the work mobilizes its resources as a physical artifact as well as on the user's interactions with the work and the interpretive strategies she develops—strategies that include physical manipulations as well as conceptual frameworks. In the broadest sense, materiality emerges from the dynamic interplay between the richness of a physically robust world and human intelligence as it crafts this physicality to create meaning" (33). The media-specific analysis takes into account the context in which a technotext exists, which includes its material form as the coded information on the computer, the artistic strategies that the writer encodes into it to access the work, and the worldview that the reader brings to the reading. Such works "that physically create fictional subjects through inscriptions also connect us as readers to the interfaces, print and electronic, that transform us by reconfiguring our interactions with their materialities. Inscribing consequential fictions, writing machines reach through the inscriptions they write and that write them to re-define what it means to write, to read, and to be human" (131).

A work that embodies Hayles's vision of the materiality of techno-texts is Stephanie Strickland's *V*, which has a dual existence—as the print text *V: WaveSon.nets/Losing L'una* (2002) and as the electronic version of this text, *V: Vniverse*, which Strickland coauthored with Cynthia Lawson. In transporting the print text into the electronic medium, Strickland and Lawson have totally reconceived the text as they make full use of the potential of the electronic medium to create a work that embodies the postmodern perspective both in its thematic content and in the artistic strategies that they have encoded into the interface. The two forms of *V* affect the reader differently, which is intimately linked to the differing materialities of these texts. The electronic *Vniverse* is not about solving a puzzle or determining a particular meaning of the text that exists outside the physical reality of the text. It is about the meaning-making process itself, and hence the materiality of the inscription technology employed to create it is an integral part of the reading experience, along with the reader's physical interaction with the text and the worldview that she brings to it. *V* in its two incarnations is a meditation on what it is to read, write, and live in postmodern times.

Cybertext, Technotext, and the Coding Metaphor

The two trends in electronic literature, cybertext and technotext, reflect the two perspectives present in contemporary culture, the cybernetic and embodied perspectives. The former reduces humans to information on the genes or texts to the play of information on the screen, and the latter emphasizes the embodied status of humans and the materiality of texts. The prevalence of cybernetic metaphors can be partly attributed to the back-and-forth movement of techno-scientific metaphors between the world of science and its cultural applications. In technical and cultural narratives humans are compared to intelligent machines or social or biological metaphors are used to describe intelligent machines. The mixing of metaphors promotes disembodied constructions of the human that are exploited by technocapitalist forces for their own purposes.

How do we appreciate the immense potential of information

technologies and developments in biotechnology without falling into the danger of thinking of humans as intelligent machines or intelligent machines as humanized neural nets? Many writers have written in empowering ways about the significance of the coding metaphor in the information age in both sciences and the arts. The code is the basic pattern whether in genetic research, musical creation, or computer programming and its infinite mutations or variations are intimately linked to its expression, whether in the human body, literature, or musical composition. Richard Powers's *Gold Bug Variations* (1992) explores the generative nature of the code, thereby illuminating two approaches to it; one way is to see it as a one-to-one correspondence between the code and the message, and the other is to look at it as generative in nature, as we see in genes in action, a musical composition in performance, and a computer program in execution.[14]

In *The Gold Bug*, Bach's musical score *Goldberg Variations*, genes, and the fifteenth-century Dutch painter de Bles's works are cipher texts that the characters are engaged in decoding. Jan O'Deigh, a reference librarian, wants to understand why Stuart Ressler, a successful microbiologist, quit the field when on the verge of making a breakthrough discovery. Franklin Todd, an art history student, wants to get insight into the life and works of the Flemish landscape artist de Bles. Jan's and Franklin's personal stories are intertwined with the narratives of Ressler and de Bles, which are linked to other texts—scientific texts, Poe's short story "The Gold Bug," and a series of quotations from different sources.

The invisible center of *The Gold Bug* is Edgar Allan Poe's short story "The Gold Bug," about Legrand's success in deciphering the secret writing on an old parchment giving directions for retrieving a hidden treasure. Ressler and Karl Ulrich, two microbiologists at Cyfer laboratory, provide two different readings of this story. Ulrich sees Poe's story as a purely mechanical exercise in cryptography to decode the secret message in order to reach the treasure. This attitude is reflected in Ulrich's approach to his own research, where deciphering the genetic code implies finding the sequence that encodes a specific biological function. Ressler's approach to Poe's story is different. For him the significance of Poe's story lies in how the code works, not the treasure to which it will lead.

Powers's *Gold Bug* presents art and science addressing similar questions though in totally different ways. Science must divide and subdivide reality in order to unravel its innermost secrets. In the quest for objectivity, the scientific results, once verified, are accepted as disembodied facts, and the context surrounding scientific discoveries is soon forgotten (Latour 1986). In genetic research, in which the scientific narratives present the genetic blueprint as if it exists outside its instantiation in material bodies or in the computer modeling of brain processes, the tendency is to equate them to actual brain processes, which are always embodied. Unlike scientific descriptions, literature and the arts strive to catch reality in its totality. Ressler, who has a great reverence for science, finds it unfulfilling for this reason. He devotes his life to Bach's music, and on his death he leaves behind a trunk full of his handwritten musical scores. In a story that he relates to Franklin about listening once again to a performance of Bach's *Goldberg Variations* in his later years, he expresses his shock at how different the piece sounds. He is "shocked to hear that it's not the same piece, not the same performance. It's radical rethinking from beginning to end, worlds slower, more variegated, richer in execution" (636). Ressler thus emphasizes that the performance of a piece is intricately linked to the occasion, which is unique and can best be presented as the performance of the same musical score with and in difference.

Powers's *Gold Bug Variations* is a complex nonlinear text with fractal enmeshed narratives revealing and reflecting variations and resemblances whose meanings can be grasped by focusing not on any single narrative in itself but rather on how these different narratives relate to one another. Powers incorporates the paradigm of nonlinear dynamics both as the theme and as the structuring principle of the novel. In highly complex nonlinear systems an underlying order exists that is not a linear order as seen in the surface patterning of individual constituents of the system but rather lies in the shift from one level to another that the system demonstrates as it proceeds toward more complexity. Recursion in a complex system implies that different levels show some invariable features along with differences. As systems become more complex, unpredictability sets in, so the final product could be new, broken away from any predeterministic logic. The very theme of the novel focuses on the uncertainty, novelty, multiple interpretations, and contingency involved in any reading, be it scientific or artistic.

Postmodern writers in general explore the plurality of worlds and how those worlds relate to one another.[15] However, it is only the critically aware writers who critique dominant technocratic narrative as they unearth other narratives and place them alongside the dominant one to create cracks and fissures through which other voices and other stories can be heard. As an assemblage of perspectives comes together in the unified representational space, different elements enter into relationship with one another. This has the potential for multiple trajectories of meaning. The contemporary topology is thus composed of discontinuity, fragmentation, multiplicity, and collage or assemblage aesthetic that is inclusive of the contradictions, differences, and ambiguities of life as it is happening rather than life as it has already happened.[16] The text is no longer a reflection of the real with a unitary meaning; instead it becomes a doorway to multiple worlds that enter into a dynamic relationship with one another. This point can be made clear by referring to Lewis Carroll's *Alice's Adventures in Wonderland* and *Through the Looking Glass*. The looking glass in Carroll's tale loses its power to reflect. As it becomes transparent, it turns into a doorway leading to a world made of pure events. In the world of the looking glass Alice encounters everything in the process of changing into something else. Similarly the disjunctions and breaks in a hypertext narrative are doorways to enter the narrative space, which is continually transforming as the narrative unfolds.

Theorizing the postmodern culture or cultural texts in terms of the performative denies supremacy to any cultural text because no text can exist apart from the living experience of these texts. Moreover, it reaffirms the embodied status of humans as well as the materiality of texts. The notions of purity and mythic origins belong to the linear temporality of the print aesthetic, which must seek claims to an origin in order to justify its end—something that has been reinforced by western colonialism, nationalism, and the scientific materialism of the preceding centuries. The linear and horizontal temporality leading to closed texts of both the east and the west is rejected in the decentered and nonhierarchical hypertext environment, where both texts and subjectivities are experienced as processual entities whose essence lies not in synchronous presence but in asynchronous realization of moments of repetition and difference. Here the past of long ago and the future of over and beyond are both joined in the contemporaneity of the present, which is an emergent quality. In this respect, an anecdote from Michael Joyce's book *Of Two Minds* comes to mind in which he relates his conversation with a philosopher. The philosopher is worried what would happen to Plato in a hypertext environment. Joyce rightly speculates that the philosopher's fear lies in the realization that as the readers trace their own paths through a platonic or nonplatonic maze they might recover their own versions of Plato, which might be very different from that of the philosopher. And I might add that they might discover their own Plato, a Plato who comes not from Greek antiquity but from the margins of Asia or Africa. The real potential of electronic media thus lies in the fact that they embody the multiplicity of postmodern experiences, in which both dominant and minority narratives coexist. Many people all over the world are turning to minority narratives to find what is missing in an increasingly consumer-oriented society. Such narratives provide a space in which nontechnocratic values and their relevance to humans can be explored. As such, they have the potential to serve as a counterdiscourse to the dominant technocapitalist narrative in which the future has already happened and all we need to feel fulfilled or empowered is another quick fix as presented by the infinite self-perpetuating mediascape of the twenty-first century.

1 DISCONTINUITY

In-between Spaces and Itineraries

POSTCOLONIAL PERSPECTIVES APTLY REPRESENT the reality of the fin de siècle, because they incorporate disjunctions and differences that were suppressed under western metanarratives of progress or social justice. In postmodern times, the comfort and stability of first principles of any kind is slipping away, while national boundaries are increasingly becoming porous. Humans and capital flow in myriad patterns in a network of relations that spans the globe. As nation-states loosen their hold on the imaginations of people in a world of transnational capitalism, the role of information technologies is crucial indeed. These technologies are having a profound impact on our literary as well as our artistic practices, creating a new space that demands its own aesthetic. This new aesthetic, which I term a hypertext aesthetic, represents the need to switch from the linear, univocal, closed, authoritative aesthetic involving passive encounters to that of the nonlinear, multivocal, open, nonhierarchical aesthetic involving active encounters.

The intertextual and interactive hypertext aesthetic is best suited to representing postcolonial cultural experience because it embodies our changed conception of language, space, and time. Language and place are no longer seen as existing in abstract space and time; rather, both exist in a dynamic interaction of history, politics, and culture. Time is no longer the linear historical time of traditional historiography, a historical time that ignored the question "Whose time is it that is being recounted?" a time that muted minority voices in a discourse based on the othering of the world. In order to escape the homogenizing and universalizing tendency of linear time, time in both postcolonial and hypertextual experience is represented as discontinuous and spatialized. The hypertextual and the postcolonial are thus part of the changing topology that maps the constantly shifting, interpenetrating, and folding relations that bodies and texts experience in an information culture. Both discourses are characterized by multivocality, multilinearity, open-endedness, active encounter, and traversal. Here I focus, on the one hand, on representational modes, that is, fragmentation–discontinuity, multiplicity–multilinearity, active traversal–active encounter in the hypertextual environment, both computer and cinematically generated. On the other hand, I explore the threads of the social, the political, and the historical that interweave the subject of representation.

Technoenthusiasts have imagined computer–human interaction in cyberspace resulting in disembodiment. However, it can be seen as a new mode of embodiment marked by moments of instantiation as well as desubstantiation. This contemporary topology is composed of cracks, in-between spaces, or gaps that do not fracture reality into this or that but instead provide multiple points of articulation with a potential for incorporating contradictions and ambiguities. Also, the in-between spaces themselves become the object of discourse as well as artistic representation. Artists, in both visual and verbal media, have felt compelled to reconfigure and rearticulate this new orientation that bodies and texts have assumed in the information culture. In her films *Reassemblage* and *Naked Spaces—Living Is Round*, Trinh T. Minh-ha uses strategies surprisingly similar to those used by hypertext fiction writers in the electronic media in order to open up new ways of seeing beyond the glass surfaces of normal vision. The technologized media, therefore, do not flatten the subject but disperse it along new lines and give it new configurations.

The hypertextual has become an "environment" or a space that demands different mappings. Jay Bolter, in *Writing Space* (1991), describes hypertext as a network of texts that allows the reader to choose any path; all paths are equally valid readings, "and in that simple fact the reader's relationship to the text changes radically. A text as a network has no univocal sense; it is a multiplicity without the imposition of a principle of domination" (25). In a hypertext environment the reader can be explorative and choose multiple paths, thereby actively participating in its unfolding. Michael Joyce (1995) makes a distinction between exploratory hypertexts, in which readers create their own paths through a body of knowledge, and constructive hypertexts, in which writers collect, shape, and act on information and create visual maps of structures that are "versions of what they are becoming, a structure for what does not yet exist" (42). He develops the idea of topological mapping while exploring the roles memory and proprioception play in hypertext reading and writing. As we experience the world in time, he writes, we remember it in space; hypertext can be seen as a "city of texts" that represents a space in which the inner space of memory and the physical space of writing come together. "Our intuition," he writes, "is that we write proprioceptively, as the child's hand does, summoning the space of memory outward. It is a dream of depth, expressed in the depth of an elastic and windowing world" (171). Joyce regards "hypertext" as "an art form [that] concerns itself with constant reconfiguration and so is a true electronic medium. Hypertext is before anything else a visual form, a complex network of signs that presents texts and images in an order that the artist has shaped but which the viewer chooses and reshapes" (206).

The narrative strategy used in the hypertextual environment lies in navigating through a body of lexias or textual segments that allows the tracing of varied paths in the midst of open possibilities. Hypertextual tracing does not aim at reaching a destination; rather, the act of tracing itself becomes the object of navigation, so that the discrete nodes are subordinated to the lines of traversal. As Harpold sees it, "Each moment of the journey-as-navigation is conditioned by the deferral that shapes its entire trajectory" (128). In hypertext fiction, series of lexias in random sequence produce a textual surface that is fragmented and discontinuous, so that no two readings of such a work are similar. In the hypertextual environment, viewers or readers do not disperse along the information superhighway but are active decoders of the path that they create in a proprioceptive act in which inside and outside coalesce in a space constituted of moments of textual embodiment and disembodiment. In modernist art, Fredric Jameson notes, the materiality of the bodies and the object world is translated into another kind of materiality of the aesthetic world so that the work evokes in the viewer the object world that it is representing. In hypertextual or postcolonial cultural productions the materiality of bodies and the object world is transformed into an aesthetic act that is intertextual, in which the text and the reader occupy the zone of the in-between of the transformation itself. The "centered" or "monadic" subject of the modernist era is thus transformed into the nomadic subject, no longer passively contemplating the artist's expression but actively reshaping it.

Trinh T. Minh-ha uses similar terminology to describe the exploratory camera work in her films about West Africa, *Reassemblage* and *Naked Spaces—Living Is Round*, which are not documentaries in the conventional sense. Her camera explores the textured lives of people and the living spaces that they inhabit. The living spaces and people are transformed into textual surfaces of multiperspectival representation. Linearly progressing images in motion are frequently replaced by multiple shots of the subject through the use of jump cuts, a technique that conveys the filmmaker's awareness of the constructed nature of cinematic representation even as it reveals the impossibility of ever capturing the "whole." And although the visual space is fractured by cinematic montage, the aural space of the music soundtrack is punctuated by long silences. Trinh's play with sound and silence can be traced to her interest in John Cage's experimental music. Cage was deeply interested in eastern philosophies. He saw a relationship between Zen emptiness and the use of silence in his musical compositions. The concept of silence evolved in his thinking: at first he viewed sound and silence as excluding each other, then he came to see them as existing together, and finally he arrived at the concept that silence or emptiness is full of sound (de Visscher, 129).

Both visual saturation and aural silence function as part of Trinh's transformative aesthetic, which produces fragmentation and difference. For Trinh, fragmentation is not a binary opposite to the whole; she feels the need to go beyond any binaries so that fragments exist on their own, and hence fragmentation implies "a way of living with differences" (Trinh 1992a, 156). She believes the term *fragmentation* can be advantageously used as denoting self-limitation, because "the self, like the work you produce, is not so much a core as a process, one finds oneself, in the context of cultural hybridity, always pushing one's questioning of oneself to the limit of what one is and what one is not. . . . Fragmentation is therefore a way of living at the borders" (156–57).

Fragmentation, then, becomes the topological mode that produces the text of becomings. As in-between spaces proliferate along multiple routes, the hypertextual subject remains inscribed in a mode of fluid transformation. From this perspective, Jameson's equation of the fragmentation and discontinuity of postmodern cultural productions with a schizophrenic subjectivity is itself nostalgic because it is rooted in a formulation of the self that is enclosed, centered, and bounded. Postmodern and postcolonial art does not locate the centered subjectivity of the artist but expresses his or her textual embodiment in a possible historical moment, always acknowledging that any artistic representation is just that—a representation, not a total, complete picture of reality. Fragmentation and discontinuity, then, comprise tools that these artists use to allow for polyphonic narrative structures.

Trinh also focuses on cinematic methods that bring out the role of the filmmaker and mark her politics of representation. Her camera work is hesitant, sudden, and unstable. She describes the exploratory movements of the camera "as a form of reflexive body writing. Its erratic and unassuming moves materialize those of the filming subject caught in a situation of trial, where the desire to capture on celluloid grows in a state of non-knowingness and with the understanding that no reality can be 'captured' without trans-forming" (Trinh 1992a, 115). Hesitance and silence express a politics and aesthetic that refuse totalization; they are the technical analogues and expressions of fragmentation and discontinuity.

In documenting West African villages, Trinh focuses on the practices of the daily life of the people. Her films transform the topographical places into topological spaces that trace the ensemble of spatializing practices of the people. The narrative trajectories traced by Trinh's camera are marked by mobile, folding, and interpenetrating relations among people, nature, and the cultural matrix of which they are a part. Trinh's films, as body writing, do not map spaces but create shifting storylines of linkages that do not crystallize into any fixed form. Michel de Certeau, in *The Practice of Everyday Life*, describes spatializing practices in terms of stories that people perform as they go around, engaged in their daily activities. Stories as spatial trajectories "traverse and organize places; they select and link them together" (115). Analyzing spatializing practices moves our attention away from structures to actions, from place to space. De Certeau compares space to the spoken word: "space is like the word when it is spoken, that is, when it is caught in the ambiguity of an actualization, transformed into a term dependent upon many different conventions, situated as the act of a present (or of a time), and modified by the transformations caused by successive contexts" (117). As a "practiced place," space, comprised of mobile elements, is fluid, and hence it lacks stability or a fixed form (117).

Whereas place is represented by a topographical map, space is represented by a topological itinerary; for de Certeau, the former involves an act of seeing and hence "the knowledge of an order of places," while the latter concerns itself with an act of going that involves "spatializing actions" (119). As two modes of experience, the itinerary centers around "a discursive series of operations," whereas the map, by colonizing space, realizes itself as "a plane projection of totalizing observations" (119). With the ascendance of scientific discourse, de Certeau notes, itineraries were slowly replaced by maps, even though the former were "the condition of [their] possibility" (120). The hypertexual medium is also composed of mobile elements; the textual body comes momentarily into existence by means of the spatial trajectory traced by the reader, who actualizes it through situating it and temporalizing it. The irreducibly multiple spatial trajectories of hypertext reading and writing transform the stable format of topographically fixed print text into itineraries of hypertext.

The immateriality of hypertext as an image display, a temporary configuration of pixels, and its capacity for instant mutation arise from the pattern-randomness dialectic that is the basis of electronic textuality. Hayles has elaborated on this pattern-randomness dialectic, describing it as the dialectic of the information age in which actual commodities of production as well as consumption are bits of information stored as electromagnetic signals of pattern and randomness. Building on Friedrich Kittler's description, in *Discourse Networks*, of fixed-type print as based on a presence–absence dialectic in which each keystroke corresponds to the letter typed, Hayles (1993) shows how computer-generated text consists of signifiers that exist "as a flexible chain of markers bound together by the arbitrary relations specified by the relevant codes" (77). The machine code, the compiler language, and the processing language are all involved in a series of operations that produce "flickering signifiers" on the video screen (70). The signifier at one level becomes signified at another, and because the relationship between them at each level is arbitrary, a global command can bring about large changes in the text (70–71). The instant mutation of the text with respect to itself, and also as it is connected to other texts, allows for the folding of relations in the topological space that is thus created. The emergent nature of the actualization of hypertextual space, which is situated in the present act and transformed by successive transformations in context, has led theorists to see the similarities between hypertext and oral literature. Walter Ong notes that electronic technology has introduced an age of "secondary orality," which is very similar to that of preliterate oral cultures in "its participatory mystique, its fostering of a communal sense, its concentration on the present moment, and even its use of formulas" (136). Because hypertext reading and writing involve active encounter and traversal, the reader and writer become integral parts of the topological space created by the interaction of multiple texts. In fact, hypertext reading and writing, like Trinh's explorative camera work, can be regarded as "reflexive body writing" of the text; the path the reader traces marks the materialization of his or her nomadic subjectivity.

The fragmentation and discontinuity that define the hypertextual environment do not lead to a fractured reading experience. In fact, the links between the nodes promote multiple narrative trajectories.

Instead of thinking of links as pointers directed at the nodes where the lexias or meaning-forming units reside, we can regard the links as positive invisible spaces with unlimited potential. These spaces lead to both disruption and continuation in the narrative stream. Carolyn Guyer and Martha Petry (1991b), who wrote their *Izme Pass* (1991a) as a fourth text based on connections within a triad of texts—Michael Joyce's *WOE,* Guyer's *Quibbling,* and Petry's unfinished hypernovel Rosary—note that the most important thing that they realized from this experience was "*how* things are connected, not connection as conceptual negative space, but connection being itself a figure against the ground of writing" (82).

The space circumscribed by the link and the node can be regarded as embodying the presence–absence dialectic in virtual space, even though at another level the two are created by the dialectic of pattern–randomness. The link thus exists as the space only in its potential, neither this nor that but as the third space that marks the site of encounter between the two nodes. The pattern–randomness dialectic enfolds in traversing lexias, because the link's referential function is constantly subverted by the disruption and discontinuity in the transforming narrative trajectory. As Harpold has written, "What you are unable to do with the link is as significant as what you might do with it" (129). If instead of assigning the function of textual communication to the link we regard it is "as a trace of the iterability of hypertextual threads, as the shape of the turn that divides them and subverts the limiting traits of context, then navigation across (or by means of) the link amounts to moving within a dilatory space whose limits can't be circumscribed" (132–33). As such, it marks the space that can be defined neither as continuity nor as discontinuity but as something that goes beyond these limiting terms and joins the nodes in the trajectory of the open text in an always emergent context.

George Landow (1993) describes the effect of electronic linking in terms of dispersal of an individual lexia into others because the physical and intellectual boundaries separating the lexias are continually crossed (53). The border crossings made possible through electronic linking lead to spatially folding relations in the open space of hypertext or culture. This has been described as leading to "the disappearance of a stable, universal context." The "denaturing" of context is also said to lead finally to a collapse of "the distinction of text and context altogether" (Hayles 1990, 272, 275). Mireille Rosello argues that hypertextual space is "an environment most likely to make the very notion [of context] irrelevant and obsolete," even going so far as to say that "it contributes to making it redundant" (133). A better conceptualization of the hypertextual space would be not to regard it as the dematerialization of the text and the consequent dissolution of context but as an always emergent context of embodiment marked by novelty and creativity. Thus the readers reading in the hypertextual environment are engaged in a "reflexive body writing" characterized by moments of textual embodiment (pattern) as well as disembodiment (randomness) as they trace their own unique path through the web of multiple texts.

Stuart Moulthrop (1994) explores the interface between information technology and culture in terms of the confrontation between the smooth and the striated, the two types of social spaces or cultural registers that Deleuze and Guattari elaborate upon in *A Thousand Plateaus*. Smooth space is characterized by becoming, uncertainty, and novelty, to which the trope of nomad or nomadology could be applied, where points of arrival or departure are subordinated to lines of flight. In the nomad's life, "the in-between has taken on all the consistency and enjoys both an autonomy and a direction of its own" (Deleuze and Guatarri 380). The definitive trope of smooth space is the rhizome, which "is composed not of units but of dimensions, or rather directions in motion"; hence, the rhizomatic space constitutes "an acentered, nonhierarchical, nonsignifying system," which is "defined solely by a circulation of states" (Deleuze and Guattari 21). The striated space, on the other hand, is the space ruled by logos and is characterized by order, causality, sequence, and hierarchy, and its distinguishing metaphor is the tree with its roots. The smooth and striated both exist; in fact, the smooth space constantly transforms into striated and the striated into smooth space. Thus the two spaces persist in a relation of difference. Moulthrop argues that hypertext and hypermedia environments represent the smooth space in which links serve as spaces of rupture that introduce openness, discontinuity, and unpredictability in reading such texts. Thus he says that "work in hypertext will involve a constant alternation between *nomos* and *logos*. We will create structures which we will then deconstruct or deterritorialize and which we will replace with new structures, passing again from smooth to striated space and starting the process anew" (316).

Trinh (1991) formulates a similar dialectic in terms of strategies of displacement, which she sees as indispensable in conveying minority experience. Displacement, she writes, creates a dynamic whereby "[each] itinerary taken, each reading constructed is at the same time active in its uniqueness and reflective in its collectivity" (23). The postcolonial writer thus favors multilinear narrative as opposed to linear factual narrative, taking

> delight in detours. Her wandering makes things such that even when Reason is given a (biblical) role, it will have to outplay its own logic. For a permanent sojourner walking barefooted on multiply de/re-territorialized land, thinking is not always knowing, and while an itinerary engaged in may first appear linearly inflexible—as Reason dictates—it is also capable of taking an abrupt turn, of making unanticipated intricate detours, playing thereby with its own straightness and likewise, outwitting the strategies of its own play. (24)

Thus, the de/re-territorialization of the trajectory and the strategies of reversal and displacement become the stylistic tools of the postcolonial writer and filmmaker. Trinh T. Minh-ha's theoretical works are surprisingly hypertextual in their arrangement and imagery. *Woman, Native, Other* and *When the Moon Waxes Red* are divided into chapters that can be read in any order. The prose is interspersed with clips from her films, images that break the continuity of the text. The chapters, interchapters, and film clips stand in the same relationship as the nodes in the hypertext connected with links. The disruption of linear progression and the lack of a single narrative thread necessitate a continual negotiation of difference.

The construction of postcolonial subjectivity is based on the fundamental assumption of the incorporation of differences, because in "borderlands" reality is "not a mere crossing from one borderline to the other or that is not merely double, but a reality that involves the crossing of an indeterminate number of borderlines, one that remains multiple in its hyphenation" (Trinh 1991, 107). In *Borderlands/La Frontera* Gloria Anzaldúa envisions the hybrid nature of the subjectivity of the people who live in borderlands based on race, gender, class, or sexual orientation in terms of the new "mestiza," who has "a tolerance for contradictions, a tolerance for ambiguity" (79). Trinh also sees the borderlines at which the hybrid subjectivity is located as constantly dissolving and reforming, revealing a place that is "always-emerging" and "always in the making." She does not want to identify with any one definition of the borderlands; instead she wants "to play with it, or to play it out like a musical score," so that identity is "not an end point in the struggle" but "more a point of departure" (Trinh 1992a, 138, 140). She notes the strategic importance of the identity claim and the importance of all three—the political, historical, and cultural—in any self-understanding. "The reflexive question asked," she writes, "is no longer: *Who* am I? but *When, where, how* am I (so and so)?" (157). Strategies of reversal, she posits, are ineffective without the strategies of displacement. Recognizing the shifting nature of cultural identity makes the notion of cultural displacement very important in postcolonial discourse. Place is no longer a topographically situated geographical location; instead, it is a topological space created through a dynamic interaction of politics, language, and culture. Trinh argues that "the notion of displacement is also a place of identity: there is no real me to return to, no whole self that synthesizes the woman, the woman of color and the writer; there are instead, diverse recognitions of self through difference, and unfinished, contingent, arbitrary closures that make possible both politics and identity" (157).

Postcolonial narratives cannot afford to be nonpolitical because of the writer's location in the matrix of power relations. Trinh (1992a) rejects theories that regard art as self-expression because they lead to the artist's privileged status based on her inner vision while obscuring the roles class, race, and gender play in the creation of the artist and the acceptance of her work. In an interview she said that the "personal in the context of my films does *not* mean an individual standpoint or the foregrounding of a self. I am not interested in using film to 'express myself,' but rather to expose the social self (and selves) which necessarily mediates the making as well as the viewing of the film" (119).

In *Naked Spaces—Living Is Round* Trinh recounts an African legend that speaks of Ogo, who rebelled against the god Amma by introducing diversity into the original unity. As punishment, Ogo is turned into a fox, who, having lost his speech, can communicate only by touching with his paws divination tables that consist of figures drawn on the smoothed sand by village diviners before sunset. The fox is lured by peanuts carefully scattered over the tables. The diviners return after sunrise and read the path traced by the fox's footprints, which have joined, encircled, or avoided the figures. The diviners' readings vary depending on the path the fox has traced during the night. The filmmaker is like Ogo, who dares to rebel against the homogenizing tendency of the dominant culture and the mainstream films that perpetuate, standardize, and normalize this homogeneity. The very rebellion implies that the filmmaker must unlearn the dominant discourse and learn new ways of communicating insights that are based on the recognition of difference. When transplanted into the Western context, this legend explains the impossibility of communicating difference in the standard linear western narrative based on dualistic thinking.

Trinh makes a distinction between territorialized and deterritorialized knowledge: the former deals with colonizing and mastering the unknown by setting the unknown as the other that must be appropriated in an attempt to make it known. And as the "sight or site" becomes known or is made visible, it becomes subject to the colonizer's grid of power and knowledge. Territorialized knowledge involves fixing people and places into stable configurations in which the interrelations among the individual constituents are already mapped out: the maps define, categorize, and immobilize the spaces in which people move. In her account of nineteenth-century European travel narratives, Mary Louise Pratt (1986) shows such a mapping of territorialized knowledge in which the normalizing, generalizing voice of the European traveler or writer "scans the prospects of the indigenous body and body politic and, in the ethnographic present, abstracts them out of the landscape that is under contention and away from the history that is being made—a history into which [indigenous people] will later be reinserted as an exploited labor pool" (145). The European presence is concealed through this discursive configuration, which

> texually splits off indigenous inhabitants from habitat. It is a configuration which, in (mis)recognition of what was materially underway or in anticipation of what was to come, verbally depopulates landscapes. Indigenous peoples are relocated in separate manners-and-customs chapters as if in textual homelands or reservations, where they are pulled out of time to be preserved, contained, studied, admired, detested, pitied, mourned. (145–46)

In postcolonial discourse, the ethnographic portraits are exposed as part of the colonial project of commercial expansion of the new frontier in order to map sites, define parameters, and depopulate the landscape so that it can be appropriated, exploited, and colonized.

Deterritorialization, then, involves uncovering the dynamic, mobile relationships among people as well as their relationship with the spaces they inhabit. The focus thus shifts from maps to itineraries and from fixing and defining through the gaze to active encounter through touch. The strategic importance of deterritorialization and reterritorialization marks the site of cultural displacement. Instead of an "authentic" portrayal of cultures, the shifting, mobile relationships of cultures coming in contact with one another become the focus of attention. In the postcolonial context, the metaphor of touch replaces the metaphor of vision, in which the gaze controls and appropriates the unknown and brings it to the realm of the known. Thus the metaphor of touch rather than that of vision is appropriate to describe postcolonial experience.

The active encounter through touch creates zones of hybrid languages and experiences that Pratt (1992) calls the "contact zone." Through the use of that term she wants "to invoke the spatial and temporal copresence of subjects previously separated by geographic and historical disjunctures, and whose trajectories now intersect" (7). The "contact" perspective, she adds, "foreground[s] the interactive, improvisational dimensions of colonial encounters so easily ignored or suppressed by diffusionist accounts of conquest and domination" (7). It also emphasizes "how subjects are constituted in and by their relations to each other" (7). Thus the interactions "among the colonizers and colonized, or travelers and 'travelees' [is] not in terms of separateness or apartheid, but in terms of copresence, interaction, interlocking understandings and practices, often within radically asymmetrical relations of power" (7).

Homi Bhabha (1995) terms the mobile zone of interaction the "enunciation of cultural difference" or the "Third Space," which is marked by "hybridity" (208). Any enunciation of cultural difference, when seen as an act that transforms even as it creates a representation, can never capture the whole of culture—not only because the act of representation is not transparent but also because an act that is always in the state of becoming cannot be fixed into any stable final formulation. Because the Third Space marks the site of encounter between the self and the other, it cannot be represented in itself. It therefore "constitutes the discursive conditions of enunciation that ensure that the meaning and symbols of culture have no primordial unity or fixity, that even the same signs can be appropriated, translated, rehistoricized and read anew" (208). Recognizing the existence of the "split space of enunciation" or the Third Space in theoretical discourse, Bhaba writes, "may open the way to conceptualizing an *inter*national culture, based not on exoticism or multi-culturalism of the diversity of cultures, but on the inscription and articulation of culture's *hybridity*." Through retrieving the third space from its invisible status, he concludes, "we will find those words with which we can speak of Ourselves and Others. And by exploring this hybridity, this 'Third Space,' we may elude the politics of polarity and emerge as the others of our selves" (209).

Bhabha's notions of the "Third Space" and "hybridity" have come under criticism. Theorists note that the concept of hybridity does not take into consideration the unequal power relations that the dominant and minority cultures embody so that any hybridity resulting from the encounter between the two is already determined by the dominant cultural norms. Furthermore, Bhabha uses "hybridity" to overcome the western binary logic, but this term easily lends itself to problematic binary oppositions. The formulation "hybrid cultures" usually leads to the conclusion that there are "nonhybrid" alternatives to postcolonial cultures, for example, western neocolonialism and third-world nationalism, which does not do justice to their "internal contradictions and differential histories" (Moore-Gilbert, 129). Bhabha's theorizing is susceptible to this critique because he uses "hybridity" and "Third Space" interchangeably in his writing. His notion of hybridity introduces a static essentialist concept that reduces the postcolonial subjectivity to a particular state of being, even as his theoretical framework emphasizes performativity. Bhabha's confusion of categories introduces contradictions into his theoretical framework. In order to allow the Third Space to be a mode of political intervention, it is necessary that the focus remain on the performativity or the process of identity formation in difference.

The active encounter of the contact zone marked by fluidity does not lend itself to linear modes of representation, so postcolonial writers have created narratives that spread out horizontally. This exfoliation defines not only the nature of textual topologies but also the spilling of texts across other genres, for example, autobiography into nonautobiography, fiction into nonfiction, or prose into poetry. Just as hypertextual narratives written by both male and female writers have been described as being similar to oral narratives in their transmutability and ephemerality, narratives by many ethnic women writers are marked by orality, which they achieve through the use of images and stories that flow and are transformed into other images and other stories, a strategy that embodies the impossibility of arriving at a definitive reading.

Because of the fluidity and transmutability characterizing the narratives of the contact zone, Trinh (1991) proposes the replacement of the central symbol of patriarchal societies—the sun, representing logos, order, clarity—with that of the moon because of its cycles and constant change, its waxing and waning. She notes the need for a dynamism of naming that does not become rigidified into commodifiable categories offered for consumption. Trinh challenges conventional modes of filmmaking in order to critique their alliance with the regimes of power that seek to turn everything into commodities of consumption. Because the mainstream cinema circulates images for consumption, which become commodities attached to a meaning, the role of the filmmaker, she argues, is to free the image from the "already attached" meanings and let it speak for itself "with its excess, its radical or unjustifiable character" (110). To bring what is invisible in an image to the surface "implies disturbing the comfort and security of stable meaning that leads to a different conception of montage, of framing and reframing in which the notions of time and of movement are redefined, while no single reading can exhaust the dimensions of the image" (110–11).

Trinh achieves hypertextual montage by shifting her attention away from action images that tell a linear story using characters that move in abstracted time and place toward letting her camera focus in a decentered way on the daily activities of the people, producing topological spaces in her films that demand the viewer's participation in their unfolding. In her films, she notes, "the relation between filmmaker, filmed subject, and film viewer becomes so tightly interdependent that the reading of the film can never be reduced to the filmmaker's intentions" (109). Trinh describes her film as a sheet of paper that she presents to the viewers; she is "responsible for what is within the boundary of the paper but [she does] not control and [does not] wish to control its folding. The viewers can fold it horizontally, obliquely, vertically; they can weave the elements to their liking and background" (109). The extreme openness of her films, with their narrative trajectories that spill in different directions, is realized through an extensive use of montage, which creates images linked through light, music, color, and sound, so that reading her films, as opposed to just seeing them, becomes a creative experience.

Cinematographic montage, of course, became a prized formal aesthetic of the cinema in the hands of D. W. Griffith (*The Gray Shadow, The Mark of Zorro,* and *The House of Hate*) and Sergei Eisenstein *(Battleship Potemkin, October)*. Although Griffith confined himself to montage in the form of parallel action in his films, Eisenstein brought it to even more complex form by exploring montage both within a single shot as well as in the entire film and by describing it in terms of conflict or "the collision of independent shots—shots even opposite to one another" (107). Within the shot he perceived montage "in the development of its intensity shattering the quadrilateral cage of the shot and exploding its conflict into montage impulses between the montage pieces" (98). In *Cinema 1* Deleuze elaborates on the nature of montage and how it works by exploring it in terms of relationships that are created as well as dissolved between the parts and the wholes through the creation of "movement-images." Commenting on Bergson's critique of cinema as cinematographic illusion, Deleuze points out that Bergson's critique was in fact directed toward primitive cinema, which tried to imitate reality in the form of action images that represented a false movement in time because the images as instantaneous sections were immobile and the time of production as well as that of projection was abstract and impersonal.

Thus the primitive cinema was like topographically fixed print text with immovable sections that followed a unilinear path, with a beginning, middle, and an end that remained the same with each viewing. However, long before print text was challenged by hypertext, the cinema evolved and overcame its own limitations "through montage, the mobile camera and the emancipation of the view point, which became separate from projection" (Deleuze 1986, 3). Through montage, Deleuze adds, an indirect image of time or the whole is created that is not a "homogenous time or a spatialised duration"; it is instead "an effective duration and time which flow from the articulation of the movement-images" (29). Exploring in greater depth the relationship between montage pieces, Deleuze refers to continuities through which parts enter into relative continuities with the sets, even as there are ruptures and discontinuities that reveal that the whole is not present. The whole, defined in terms of relations, appears at another level that cannot be reduced to discontinuities or continuities; rather, it "appears in the dimension of a duration which changes and never ceases to change. It appears in *false continuities* . . . as an essential pole of the cinema" (27–28). Further elaborating on the notion of false continuity, Deleuze adds that it "is neither a connection of continuity, nor a rupture of a discontinuity in the connection. False continuity is in its own right a dimension of the Open, which escapes sets and their parts. It realizes the other power of the out-of-field, this elsewhere or this empty zone, this 'white on white which is impossible to film'" (28).

Projected onto the topology of hypertext, Deleuze's "false continuity" in cinematographic montage can be seen as the space circumscribed by the hypertextual link. This space, neither a mark of continuity nor a site of rupture, escapes the defined boundaries of nodes, marking the site of the in-between even as it is reunited with them at another level that belongs to the dimension of the whole, but a whole that is open. "False continuity" is also similar to Bhabha's "Third Space" or the "split space of enunciation" that marks the site of cultural encounter. Again, it can be compared to Trinh's "negative space," which she attempts to capture in her films. Trinh's negative space is neither the space that the subject of representation occupies nor the field around it; it is "rather the space that makes both composition and framing possible, that characterizes the way an image breathes" (Trinh 1992a, 142). Trinh compares this negative space to the Zen void, which reveals the image in its multiple differences.

Trinh created movement-images in *Reassemblage* primarily through fixed shots. The duration and intensity of a shot make the image subversive; the gaze of the viewer encounters the pensive image of the object and "sees it [as] an object that speaks" (Trinh 1991, 115). Such an image does not lend itself to consumption, Trinh observes, because it speaks for itself, and in that it is a challenge to the fundamental assumptions of the mainstream cinema. It provokes viewers to question and think instead of drawing them into a world of simulacrum represented on the screen. Trinh critiques filmic conventions because these are too limiting to portray the heterogeneous experiences of life. In her films the images, as instants frozen in time, are transformed through mobile relationships between the image and the frame, the image and the image through jump cuts, light and darkness, the framed or reframed image, and the soundtrack, which is a succession of music, songs, voice-overs, sounds from nature, conversations, and silences, sometimes with overlaps and superimpositions.

As part of the movement-images, images lose their contours, attain depth, and are united in the indirect time-image or duration of the film, which brings about a qualitative transformation in the experience of the film. In a process of continual feedback, the movement relates the framed images to an open duration, and this, in turn, opens up the images. The assemblage of diversely framed images works in a polarity that oscillates between the interval that defines the immediate relations of the image and the spiral or circle of time that puts it in relation to the duration of the film. Both *Reassemblage* and *Naked Spaces* end in a way that stimulates a qualitative transformation in the viewer's reading of the film. Both films begin and end with the same music or dance sequence and thus describe a circle or rather a spiral of time, so that the experience of the end sequence is qualitatively different from that of the beginning sequence.

Trinh (1992a) says *Reassemblage* was realized "as a desire not to simply mean" but rather "to expose the transformations that occurred with the attempt to materialize on film and between the frames the impossible experience of 'what' constituted Senegalese cultures" (113). *Reassemblage* can be seen as a critique of ethnographic films that collect, preserve, and display cultures as museum artifacts. Through her critique she questions the concept of "authenticity" as well as the "objectivity" of ethnographic endeavor. She compares anthropologists to fishermen who locate themselves as observers in alien cultures and then, in the name of objectivity, cast a net (their theoretical framework) to capture the culture they are observing. Trinh is acutely aware of herself as part of the net in which she is already caught as she does her catching. In her filmmaking there is no fisherman, only the net. As she focuses her camera on a woman, she watches "her through the lens, [she] look[s] at her becoming [her] becoming [hers]/Entering into the only reality of signs where [she herself is] a sign" (Trinh 1992c, 101).

Both *Reassemblage* and *Naked Spaces—Living Is Round* focus on people and the living spaces they inhabit. Trinh (1992a) describes the dwelling as an integral part of the landscape, both social and natural, so that the earth and the sky, divinities and mortals, life and death come together in her filmic representation. She says:

Dwelling is both material and immaterial; it invites volume and shape as well as it reflects a cosmology and a way of living creativity. . . . To deal with architecture is to deal with the notion of light in space. To deal with the notion of light in space is to deal with color, and to deal with color is to deal with music, because the question of light in film is also, among others, a question of timing and rhythm. Such mutual accord of elements of daily existence is particularly striking in the built environments filmed and the way these materialize the multiple oneness of life. (120)

Trinh makes frequent use of deformed pans in order to depict spatial interrelationships that the camera would not otherwise capture. Rejecting the use of artificial light to lighten the indoor spaces, she films instead in natural indoor light and thereby reveals the beauty of these spaces. The play of light and darkness gives depth to the space. Thus, she writes, "If light be called the life blood of a space, darkness could be called its soul" (1992b, 31).

Trinh (1992a) points out that the conceptual basis of *Naked Spaces* lies in the experience of momentary blindness that she frequently had in Africa when stepping into a dwelling from the broad daylight outside. She turns this experience into a metaphor for describing any shift in reality that requires readjusting and refocusing our vision. Thus, to "move inside oneself, one has to be willing to go intermittently blind. Similarly, to move toward other people, one has to accept to take the jump and move ahead blindly at certain moments of inquiry" (119). In both films Trinh represents this experience by focusing her camera on openings in the dwelling walls, looking from within through doors that open onto luminous light outside and sometimes looking inside from outside through doors or openings in the walls. There are also long moments of seeing through gaps, cracks, and in-between spaces. Trinh thus underscores not only the framed nature of her vision but also the need to take a blind leap, beyond one's beliefs or cultural assumptions, to understand the other.

Trinh's use of black screens in *Reassemblage* seems to be a reminder to the viewer that the images presented are not just to be seen but also to be read. A black screen reflects the need to have momentary blindness or emptiness in order to enter the reality of the other. At another level it marks the interval between two moments of embodiment—an interval that is stretched out and marks the site of encounter and depicts the space of traversal that connects the embodied images of the preceding and the following frames to the open duration of the film. In this respect, black screens are blank or empty visualizations of "false continuities" of the movement-images and could be compared to the emptiness of Zen Buddhism. This is not an emptiness of nothingness but that of presence, constituted not of essences but of relationships. The black screens of *Reassemblage* can also be seen as visualizations of hypertextual links that are present here in their absence as the dilatory space between the full frames, marking the site of active traversal.

Aside from black screens, Trinh's camera also focuses on empty spaces—the interiors or exteriors of dwellings or occasionally landscapes emptied of people. Deleuze (1989) describes Ozu, the Japanese filmmaker, as the inventor of empty spaces. Ozu's films, he writes, deal with ordinary rather than extraordinary moments in the lives of ordinary people, raising the banality of the everyday to moments of pure contemplation. The emptied spaces, devoid of occupants, deserted exteriors, or landscapes in nature, assume an autonomy and "reach the absolute, as instances of pure contemplation, and immediately bring about the identity of the mental and the physical, the real and the imaginary, the subject and the object, the world and the I" (16). According to Deleuze, Ozu also makes a skillful use of still lifes that are "defined by the presence and composition of objects which are wrapped up in themselves or become their own container" (16). As two modes of contemplation, "empty spaces, interiors or exteriors, constitute purely optical (and sound) situations, [and] still lifes are the reverse, the correlate" (17). Deleuze thus sees the basic distinction between the empty spaces and filled spaces as representing the emptiness–fullness dialectic in both Japanese and Chinese philosophy.

Trinh's films show a similar dialectic at work; her camera gazes at empty spaces, lingeringly moving to survey these spaces, producing an almost tactile sensation. The empty spaces are sometimes dark interiors with shadows pierced by bright luminous sunlight entering through openings in the walls or in the ceiling. At other times the empty spaces are bright exteriors of the dwellings or even landscapes. The play of light and darkness is pushed to the extreme so that objects are revealed in their emptiness and space is exposed in its unrealized possibilities. One can sense that "floating around in these dark spaces is the subtle smell of clay, earth and straw" (Trinh 1992b, 31). The white luminous light reveals objects and spaces in their full potential: "An act of light lets day in night / Makes far nearer and near farther" (29). An act of light allows one to see beyond the surfaces and experience the images as joined in the open duration of the film, which connects the interval of time marked by the empty screen with the circle or spiral of time created by the movement-images of the full screens.

From the empty spaces, interior or exterior, a sudden cut takes the viewer to the same or different spaces inhabited by people; sometimes the frame switches to the one inhabited by people. Transitions, subtle or sudden, from one form to another are quite frequent. In *Naked Spaces* a figure walking inside through a luminous doorway dematerializes into a shadow. Here Trinh seems to be playing with the interpenetration of two forms of experience—instantiation and desubstantiation, light and darkness, emptiness and presence, real and imaginary, subject and object—in which one interpenetrates the other. The emptiness–presence dialectic in Trinh's films is very similar to the textual embodiment–disembodiment that occurs in the hypertextual environment when the reader traces a narrative trajectory through multiple texts. Trinh, however, illustrates how emptiness itself is marked with presence, which she describes as the "negative space" of her films that makes the images breathe.

Trinh's treatment of sound and silence is yet another form of the dialectic I described earlier. The soundtracks in both films are punctuated with long silences; although sometimes the interruption marks a switch to a different scene, at other times the music stops abruptly even though the scene continues uninterruptedly. For example, there are sections in *Naked Spaces* in which the music suddenly stops while dancers continue dancing. Silence and sound, though aurally presented as excluding one another, are shown to be present together at another level: "Sounds are bubbles on the surface of silence," Trinh writes (1992b, 4). Just as emptiness is permeated with presence, so is silence full of sounds—as is visually displayed in the movements of the dancers dancing to a rhythm that the viewer cannot hear.

In Trinh's films the blanking of visual and aural space is done intentionally to provoke the viewer to think. In contrast to ethnographic films, which are circulated as "authentic" representations of different cultures, her films refrain from claims of "authenticity" while focusing on bringing out the multiplicity of the cultural matrix. Through an artful handling of light and darkness as well as the use of colors and music, Trinh represents a cross-section of the culture that is at the same time "clear, simple," and "irreducibly complex in its simplicity" (Trinh 1992b, 39).

Paul A. Harris's fractal conceptual model can be effectively used to illustrate Trinh's representation of a culture whose multidimensionality can be brought out only by the hypertextual handling of time and space. Even though fractal objects are discrete points, the shapes they produce are liquid, flowing, transforming into self-similar shapes with different iterations. The inside of the fractal shows an interlacing and stretching of space that exhibits a multivalent depth dimension. In this model time no longer corresponds to a linear movement; time is spatialized in a discontinuous form and appears as "the dynamic immanent to spatial trajectories" (189). Harris uses the spatially unfolding image of fractal time as a paradigm to read different spatial sites constituted by a bundle of narrative trajectories that intersect at different points and hence cannot be reduced to a single narrative. The "processual" subject is "the product of a differential, a difference located at the seam dividing the purely discrete from the smoothly continuous," and memory "reroutes seemingly discrete experiences or impressions into recursively connected loops. The incomputable, unknowable totality of possible configurations that comprises one's life is like a sprawling fractal, and one's own self-knowledge and the events in one's life take shape as routes one may follow through the fractal" (190). As the subject exfoliates along multiple routes, the relationship of language to space also undergoes a transformation. In spatialized time language is reduced to "a sequence of incursions of discourse on space" (188). Harris describes the text as a matrix that brings together disparate components across fractal dimensions "enabling reader, text and cultural context to combine within an encompassing ecology" (193). The boundaries between reader, text, and cultural context "interpenetrate and fold through one another in complex ways unrepresentable in conventional (Euclidean) spatial terms" (193).

Through her films Trinh attempts to capture just this interpenetration of reader, text, and cultural matrix so that the filming as well as the filmic subject emerge as processual entities whose contours transform as different components are seen in different combinations and configurations. The subject as presence or unitary subjectivity is replaced by the subject as nonunitary or provisional or processual. In *Naked Spaces—Living Is Round,* subjectivity is embodied through three female voice-overs. One quotes African writers and the villagers' sayings, the second follows western rationalist logic, and the third relates personal stories; sometimes the three overlap. Both language and subject are thus dispersed across a multidimensional space. Trinh spatializes time by using her camera to explore the spaces in such a way that it brings out their multivalent depth dimension. Time is no longer a sequential narration of events; rather, it is discontinuous and transformed into spatial trajectories that expose thick layers of time—"A sense of time, not only of hours and days, / but also of decades and centuries / A sense of space as light and void" (Trinh 1992b, 9). The six villages that she filmed in *Naked Spaces* are in Senegal, Mauritania, Togo, Mali, Burkina Faso, and Benin. Senegal appears at both the beginning and the end of the film. If the seven filmic sites are seen as seven temporal levels in the form of visual recursive operation, Trinh's aim seems to have been to reveal the self-embedded layers of multidimensional space of the cultural matrix. The recursive movement appears not only in the visual images and aural sounds but also in the verbal text that accompanies them. The fragmentation spatializes the linear time-bound flow of events, and instead of one flow of linearly sequenced events we have multiple self-embedded surfaces that flow into one another. The recursive loops appear in the spiraling trajectory of the narrative as well as the visual and aural space. The camera exploratively lingers for a moment too long on the objects as if in dialogue with them, letting them speak for themselves, bringing their richly textured, interlaced spatial trajectories to the surface. Trinh admits that the recursive loops in the film were used not simply to fragment or emphasize but to convey the rhythms and patterns she experienced while she was filming in different villages. The repetitive patterns are "not just the automatic reproduction of the same but rather the production of the same with and in differences" (Trinh 1992a, 114).

Trinh's use of colors—shades of red, orange, and yellow—is meant to forge flowing indeterminate relationships in an assemblage of images within the frame that coexist with the temporal order determined by the movement from frame to frame. Red is presented as "a warm limitless color that often acts as a sign of life" (Trinh 1992b, 21). Because color "is first and foremost a sensation" (11), it makes the images vibrant and suffuses them with emotional tones, making them come alive. The play of light also marks the movement of time as the day transforms into evening, the evening into dawn, and the dawn into full day. The transitoriness of life, signified by the reference to "houses and humans [as] both made of small balls of earth" (9), is set in the broader framework of the unfolding of history. In another set of visual and verbal images, women's historical location in the past and present come together:

A long wail tore through the air
Blue veiled figures
She sailed down the alley
Her indigo-blue garment
Flowing behind her. (13)

This verbal as well as visual image is followed by another that adds layers of time to it:

As if for centuries
She sat there
Instinctively veiling her face as the men came in
Unveiling it as soon as they left. (13)

Images of nature—the sun, the earth, the moon—flow into the images of living spaces, creating a fluid experience in which things named are in the following instant unnamed:

Blue like an orange
And orange becomes blue
Earth becomes Sun
Sun becomes Water
Water becomes Sky
And blue becomes orange like the Earth. (38)

It is not the naming that finally has any significance but the way it is connected to other named and subsequently unnamed images. The natural imagery merges with the images of the inhabitants of the living spaces and the activities that they perform: "It is by way of the 'house hole' that the rays of the noon Sun enter into the house to look at the family and speak with them. Family eats around it. The food cooked and spilled while eating are so much offered to the Sun. Women give birth under it to secure the Sun's blessings" (31).

In Trinh's style of filmmaking, the sounds, the images, the colors, and the music flow into one another. Music is one of the links in her films that connects the images to one another. The three voice-overs in *Naked Spaces* reflect on the role of music in bringing a sense of joy into the lives of people who respond to music through creative movement—dance. Music and dance go together. As this theme is developed at the higher, more complex, recursive level, music is described as the life force of the people, sustaining nature and people alike. In the village in Togo, we find further elaboration of the theme of music. As the film returns to Senegal in the final section, the themes developed through the exploration of sound, movement, light, color, and music once again reappear, but at a higher and more complex recursive level. Music is now described as the intermediary between darkness and light: "Music rests on accord between darkness and light" (43). The way this film tries to deal with polarities throughout is through the fluidity of images, the transformation of one into another, in which meaning lies not in the one or the other but in a space that goes beyond both.

John Cage's musical compositions for Merce Cunningham's dance choreography have led to avant-garde theories of dance and music in which dance is not seen as an expression of music, but music and dance are seen as two different art forms that could be co-performed without one expressing the other. This philosophy becomes the basis of Trinh's use of dance and music, which are presented as coexisting in their difference. In *Naked Spaces* we hear the following:

> Dance and music form a dialogue between movement and sound. One who hears the music understands it with a dance. The dancers do not imitate or express the music heard; they converse with it and dance to the gaps in it. Both marked and unmarked beats. A different beat, one that is not there, one that you add because you feel it and fit it in. Your own beat. Your own move. Your own reading. (45)

In a further elaboration, music is described as tied to movement, dance, and speech so that "the listener becomes a co-performer" (23), just as the viewer of Trinh's films becomes the co-creator in readings that have the potential for multiple trajectories.

The postcolonial and the hypertextual represent two manifestations of the topology of postmodern information culture, in which grand narratives are being replaced by local narratives and local knowledges. Trinh's films show that the hypertextual is the representational space through which the postcolonial can work most effectively. It is postcolonial discourse that brings out the politics of embodiment and shows us most clearly that bodies do not exist in transparent space. Technology might claim to have made possible a clean virtual space in which categories of race, gender, and class are said to be irrelevant and in which humans can experience the freedom of total disembodiment. We know better. Because humans are half of the interaction of the real with the virtual, if our society is not changed at a fundamental level, no leaps into virtual space can bring us freedom from the inequality and injustices of the social reality. Early hypertext theorists' exaggerated claims about the democratic potential of hypertext have not been informed by politics. Communication in itself, as Henri Lefebvre points out, cannot bring about a revolutionary transformation of society (29). Navigating in the "city of texts" can be liberating when the sites are approached not in the spirit of possession and control but in the spirit of seeing them as the sites of active encounter that is marked by a self-awareness of one's positioning. Changes in our technologies of production are intricately connected to changes in the technologies of signification, which in turn are tied to the modes of consumption and those of embodiment (Hayles 1993, 69). Postcolonial discourse, by placing bodies at center stage of postmodern topology, resists a position that promotes disembodiment. In recognizing the site of the personal, the social, and the political as the locus of struggle, postcolonial discourse acknowledges the necessity of locating the embodied body in a web of power relations, even as postmodern discourse wants to disperse it in the virtual domain. This latter view is reminiscent of the "death-of–the-subject" ideology promoted by male theorists that feminists find so problematic because it makes the need for or possibility of political action a virtual impossibility.

2   FRAGMENTATION

Gender and Performance

THE EARLY HYPERTEXTS WERE MORE LIKE PUZZLES that readers
were given to solve in order to make sense of the text in its entirety.
In the hands of writers like Judy Malloy and Shelley Jackson, however, fragmentation became a self-reflexive tool for considering
women's artistic subjectivity. Malloy's *its name was Penelope* (1993)
and Jackson's *Patchwork Girl* (1995) are poetic meditations on bringing to the surface the absence of or gaps in women's literary history
and the need to forge a women's tradition to constitute their artistic subjectivity. The hypertextual format allowed them to create a
web of interlinked text that the reader can access in multiple ways.
Each reading is different because it involves a different sequence of
textual segments and hence a different system of links that exposes
many different types of spatial relations that crystallize into multiple
reading trajectories.

Both Malloy and Jackson reflect on the psychological and sociocultural forces that go into shaping women's artistic subjectivity,
especially as they relate to women's literary history. Malloy's *its name
was Penelope* presents the difficulties experienced by women who
were raised on classics that present women as homemakers and
men as adventurers. These women felt compelled to create their
artistic vision on the periphery as they looked through cracks and
peepholes to find alternatives to the dominant constructions of the
patriarchal culture, which were oppressive to their self-expression.
Jackson's *Patchwork Girl,* on the other hand, is the story of encounters between Shelley Jackson, an aspiring writer attempting to find

her voice, and Mary Shelley, the writer of *Frankenstein*, and countless other forgotten women storytellers. In exploring subjectivity through the process of reinscription of history, Jackson unravels crisscrossing threads of connections that have been historically buried under the homogenizing tendency of the dominant discourse. Though both Malloy and Jackson approach the question of women's artistic subjectivity from different perspectives, their objective is the same—to feel empowered as woman artists.

Judy Malloy's *its name was Penelope*

Malloy's *its name was Penelope* is narrated by a forty-three-year-old photographer named Anne Mitchell. Anne's narrative appears in three main segments—"Dawn," "Sea," and "Song." The middle segment is divided into four sections: "a gathering of shades," "that far off island," "fine work and wide across," and "rock and a hard place." Each part consists of a series of verbal units, or lexias, which can be read in a sequence determined by the computer's pseudo-random number generator. The reader can jump from one section to another or use the default command to read the lexia in each section. This ensures that each reading is different, because each involves a different sequence of lexias or narrative segments and hence a different system of links that exposes many different types of spatial relations. These spatial relations then create a temporal tissue as they generate an assemblage of memories.

Malloy (1993) has noted that she wanted her work "to be the writing equivalent of the captured photographic moment," so she "utilizes the light/dark contrasts of photorealist painting" (13). The hypertextual format allows for individual memories to be presented in the manner of a photograph in a photo album, "so that the work is like a pack of small paintings or photographs that the computer continuously shuffles." The simple interface brings the reader and the narrator together so that the reader can step into Anne's mind and experience her direct vision in the form of snapshot memories (10). Even though each memory as it appears on the screen is a fragment, it is spatially connected to other memories. The narrator's distinctly visual memory photos create a hypertextual collage very similar to the cinematographic montages of Trinh's films, which lead to fragmentation and discontinuity while simultaneously opening spaces for multiple readings.

As the reader goes through screen after screen, a few images keep recurring in different forms; each iteration of an image touches off a recursive dynamic in the reader's memory. One such image is that of a boat named *Penelope*, which Anne launches in a small tidal pool as a child. The bright-blue boat has two sails that her "mother has sewn from a torn sheet," pointing to the matriarchal contribution to Anne's artistic subjectivity. As the image is repeated, the reader sees Anne launching other sailboats, even chasing after a small boat with her camera focused on it. The concrete action of launching a boat and trying to capture the image of the boat coalesce in the realization that she cannot capture reality without transforming it. Any totalizing description remains forever elusive.

Malloy uses the fluidity of the hypertextual medium to create a poetic text, which, in spite of its fragmentation and discontinuity, leads to a very satisfying reading experience because it allows the reader greater creativity as to the form the reading will take. Malloy's intentional use of a simple interface allows the reader direct access to Anne's mind without turning the narrative into a puzzle to be solved. Even though the memory photos appear as inscriptions of language on a textual surface without depth, tracing a path through them draws the reader into an intricate dialogue with the text. It turns the flat representational surface into a self-embedded spatial domain out of which emerges the nomadic subjectivity of the narrator.

In Malloy's text, the visual is transformed into the verbal. The border between text and image dissolves, and image becomes text. Malloy installs photographic conventions in the verbal medium, which is facilitated by the lexia format of the hypertext, only to subvert them to reveal their unacknowledged politics. In the first segment, "Dawn," Anne's childhood self is represented in the form of photos that crystallize distinct moments of her life. In the second segment, "Sea," the nature of photographic images, usually considered transparent, is subtly revealed to be a construction. Here the reader sees Anne engaged in the art of photography; she takes pictures in a way that suggests her awareness of her freedom to choose a subject and the way she wants to represent it, even as it is subject to the specificities of her gender, race, and class.

The epigraph in the subsection "fine work and wide across," taken from Homer's *Odyssey*, reads:

> first a close-grained web
> I had the happy thought to set up weaving
> on my big loom in hall. I said, that day:
> "Young men—my suitors, now my lord is dead,
> Let me finish my weaving before I marry,
> or else my thread will have been spun in vain."

On the same screen we read about Anne's work:

> The work I am making will be woven of twenty strips which
> I call
> tapes. Each tape is five feet long made up of color Xeroxes
> of photographs
> taken in one situation or place like
> Macy's Department store the week before
> Christmas, or The San Francisco subway, called BART, at
> 5:00 on
> a Friday, or,
> Sproul Plaza at UC Berkeley on a warm spring day,
> or TV newscasters as they appear in the evening on
> my TV monitor. *(its name was Penelope)*

Penelope's weaving centers around the fate of her "lord"; it is only through delay and explanations that she can engage in any weaving. As a ruse to defer her remarriage, it is also a commentary about the need to engage in such a stratagem because of her historical location. Also, Penelope's weaving is important not in itself but only as a means to escape her fate. Her act of weaving during the day what she unweaves at night constantly maintains the deferral of the text's end and also brings to light the emptiness that is at the heart of her project. Penelope's weaving of the web on the loom does not center on the nature of representation; rather, it foregrounds her gendered position as the crafter of a representation imbued with an ambivalent status.

Even though Anne's artwork is not a stratagem centered on her "lord," she too feels the need to stay away from thoughts of a husband and children in order to pursue her art. Malloy's narrative strategy of juxtaposing two different texts on the same screen thus highlights the gendered location of Anne's artistic subjectivity as well as its historical context. Penelope's loom exfoliates in Anne's photographs into subway systems, department stores, and nomadic routes in the cityscape. In the age of mechanical reproduction, her xerox photomontages reflect the media-dominated consumer culture of contemporary society. The interweaving of the narrative threads thus works at two levels. On the first level it brings into sharper focus the patriarchal arrangement whereby women like Penelope have historically stayed home weaving and unweaving, which means engaged in mind-numbing domestic labor, while men have gone out into the world seeking adventures. The second level reveals the politics of representation as well as historical, social, and political factors that go into the production of cultural artifacts.

Anne's photomontage is made of color photocopies of her photographs of scenes from Macy's department store, the San Francisco subway, and Sproul Plaza, which are combined with her pictures of the newscasters whom she photographs as they appear on her television screen. The photocopies of copies of televised mass media images underscore the production and reproduction of images while at the same time revealing their constructed nature. Because in a mass media culture the image is exploited for its persuasive power, representation itself becomes the subject of Anne's photomontages. Her appropriation and recontextualization of powerful images from the mass media subversively exposes their politics and leads to a different experience of these images. Anne is not in search of authenticity or an aura that has disappeared due to the very nature of the photographic medium with its power of reproduction. As Walter Benjamin notes, when the criteria for authenticity can no longer be applied to a work of art, its function as well as its grounding undergoes a drastic transformation, because art in the age of mechanical reproduction is based on politics. We need to qualify Benjamin's statement, because not all art in the age of mechanical reproduction is based on the politics of resistance.

In the contemporary era, the mediated nature of representation is an accepted fact. Any form of representation is a cultural construction, and as Annette Kuhn notes: "Photographs, far from merely reproducing a pre-existing world, constitute a highly coded discourse which, among other things, constructs whatever is in the image as object of consumption—consumption by looking, as well as often quite literally by purchase" (19). While critiquing the Baudrillardian notion of hyperreality, Hutcheon (1989) argues that in theory-informed postmodern art it "is not that representation now dominates or effaces the referent, but rather that it now self-consciously acknowledges its existence as representation—that is, as interpreting (indeed as creating) its referent, not as offering direct and immediate access to it" (34). Using highly coded visual discourse, these artists put under scrutiny both the conditions of "production" as well as "reception" of their art, thereby offering a re-visioning of the appropriated images— "a second seeing, through double vision, wearing the spectacles of irony" (Hutcheon 123).

Malloy emphasizes the framed nature of Anne's vision in memory photos that show her looking at the world through the camera lens. Anne's peripheral vision is shaped by seeing through the branches of trees, through clear water, through peepholes, through gaps in fences, or through cracks of doors, just as her reading is a reading of in-between spaces. One of the memory photos in "Dawn" is about Anne's father's reading the *Odyssey* to her and her brother. As she listens to her father while sitting on the arm of his chair, she silently reads the passages that he skips over. Not only do the titles of Malloy's work allude to Homer's epic; different sections begin with epigraphs from that epic. If the epigraphs stand for the sections that the father reads to his children, the textual spaces in between epigraphs tell the story of Anne, a photographer, who has faith and courage in her personal vision and embarks on her journey to become an artist in a patriarchal culture. Just as the narrator focuses on the gaps in the father's telling of the story, she seems to invite the reader to use the same strategy to focus on the cracks in her story. The juxtaposition of visual-verbal images creates fissures or gaps in both her photocopy-photo artwork and the memory photos of the lexias. The multiple readings of the text finally exist not so much in what the lexias say but rather in the relations they forge with one another. These relations come into existence and dissolve with each reading and unfold into different versions of the text. In Anne Mitchell's fragmentary memoryscape, the male text is reduced to a few epigraphs, whereas the female text exfoliates outward, spilling over the boundaries in multiple directions that reveal to the reader the significance of the social, the political, and the historical in any artistic endeavor.

Shelley Jackson's *Patchwork Girl*

Shelley Jackson's *Patchwork Girl* (1995) can be regarded as a transition text between the early hypertexts like Malloy's *its name was Penelope* with their exclusive focus on links and the later hypertexts in which the design of the interface became an important part of the reading experience. The interface in this work comes with a tree map, a storyspace map, and an outline of the story that provide multiple pathways for accessing the work. Jackson's text thus exhibits elements that have become an integral part of the second-generation hypertexts whose purpose is to map a navigational space and encode it in the interface design.

Jackson's confrontation with Mary Shelley in *Patchwork Girl* reflects her anxiety of authorship: the Miltonic questions "Who am I?" "Where am I?" and "Whence am I?" that the monster in *Frankenstein* asks to understand his place in the world become the thematic focus of her text. The title, *Patchwork Girl* or A Modern Monster by Mary/ Shelley and Herself, alludes to this encounter as Jackson plays on the name Shelley, which she has in common with her foremother. In exploring her literary history, Jackson must confront the historical labeling of women's creativity as monstrous, which makes Mary Shelley refer to *Frankenstein* as "my hideous progeny" (Shelley 1992, 23). Whereas Shelley's narrative ends with the exit of the male monster out the window onto an ice raft as he is "borne away by the waves, and lost in darkness and distance" (Jackson), Jackson's narrative begins with the rebirth and metamorphosis of the female monster that is made into an empowering symbol of the female artistic subjectivity.

Feminist critics regard the narrative of the monster, a female in disguise, as the heart of Shelley's *Frankenstein*. Gilbert and Gubar see in it a mock rewriting of Milton's *Paradise Lost* and describe it as a narrative about the fall of woman into gender. Thus "the monster's narrative is a philosophical meditation on what it means to be born without a 'soul' or a history, as well as an exploration of what it feels like to be a 'filthy mass that move[s] and talk[s],' a thing, an other, a creature of the second sex" (235). Jackson's text opens Shelley's text from within and rewrites its innermost core, which is constituted of the female voice struggling to understand both its rejection by its creator and its outcast status as the outsider and the other. It subverts the inside–outside distinction of the male discourse, which turns difference into otherness that is assigned a secondary status. The text exfoliates outward and makes difference and multiplicity the basis of identity and politics. As a result, whereas Shelley's monster feels alone, ugly, and disconnected from society as it is without a (his)story, Jackson's monster experiences her connection to (her) story and revels in her monstrosity, her multiplicity, and her difference. The representation of the female artistic subjectivity as monstrous becomes synonymous with the expression of difference that refuses to disguise itself or be suppressed by the male tradition.

Patchwork Girl is capable of countless mutations. The expressive representational surface that the electronic medium provides becomes a part of Jackson's storytelling. A tree map, a Storyspace map, a chart overview, an outline, and drawings of the female body all contribute to the reading experience. The tree map—a patchwork arrangement of brightly colored horizontal and vertical bars of varying lengths, shapes, and colors—itself becomes the focus of aesthetic attention. Underneath the electronic patchwork arrangement hides the textual patchwork that readers must sort out and reassemble to create a coherent whole. One of the lexias or verbal segments under "graveyard" tells readers that if they want to create a coherent whole out of the text, they will have to sew the pieces together.

Jackson's narrative, beginning where Shelley's ends, reveals through a fractured surface what Shelley's conceals in an elaborate system of framing and reframing. Shelley disguises the authorship of her text by framing the narrative with letters from Walton to his sister, followed by the narrative of Frankenstein and the innermost narrative of the monster. Jackson's text, on the other hand, contains multiple narrative folds that do not disguise the female creator of the text but rather reveal her connections to women in time and out of time. The drawing of the female body, titled "her," the opening screen of *Patchwork Girl*, holds the disparate parts of the text together as an invisible patchwork of relations through a complex system of mirrored reflections as well as refractions. The female body serves as the doorway to the Storyspace map, which shows a very symmetrical structure in this very asymmetrical text.

The map corresponds to the superimposed figure of the female body (a mirror image of the reader) with which Jackson's text opens. The top of the map (head) contains the figure of the whole seamed female body. The middle part of the text (trunk) narrates the rebirth and metamorphosis of the female monster from Mary Shelley's *Frankenstein*. The bottom of the map (legs), "crazy quilt," rewrites the male monster's narrative from *Frankenstein* as he is given a new voice that emerges through the gaps created through the juxtaposition of fragments from Mary Shelley's *Frankenstein* and L. Frank Baum's *Patchwork Girl of Oz*. The right side (the reader's left side) of the map is the modern monster's narrative, whereas the left side narrates the constitution of her body out of the fragments of the bodies of women from the past. Both the body of the text and the text of the body come together in Shelley Jackson's conception of *Patchwork Girl*.

Aside from the structural correspondence of Jackson's text to the female body, the narrative can also be explored in terms of four folds. Each fold has two layers, the figure of the female body as the outside layer and the textual narrative as the inner layer. The text of the body and the body of the text become interchangeable. Starting from the inside, the innermost fold is "hercut," with its textual narrative "crazy quilt." Encircling "hercut" are "hercut2/journal" and "hercut3/story," each constituting half of the second fold. Moving outward, the third fold is "phrenology / body of the text" and "hercut4/graveyard," each again making up half of the fold. The three narrative folds are held together by the opening screen "her," which displays the figure of the seamed whole female body. The outermost fold "her" and the innermost "hercut" are reverse images of each other. If "her" is the figure of the seamed whole female body, "hercut" represents the fragmented female body, and the corresponding textual fragments under it, called "crazy quilt," show how each continuous textual fragment is in fact an aggregate of many texts (revealed in "notes" attached to the "crazy quilt") that coalesce to give rise to a new text whose meaning lies not in any single constituent but in emerging interrelations.

The juxtaposed fragments in "crazy quilt" are primarily from Shelley's *Frankenstein* and Baum's *Patchwork Girl of Oz*. In Jackson's text, Baum's Patchwork Girl and the male monster from Shelley's *Frankenstein* become interchangeable. This section ends with a textual fragment, "but I am glad," which not only restores the female voice to the male monster but also gives a lineage to Patchwork Girl. By juxtaposing different quotations from the two texts and by filling in the in-between spaces in each lexia, Jackson is able to subvert Patchwork Girl's freakishness and the monster's monstrous self-image. The attributes that were used to brand them as freaks or outcasts become, in fact, sources of their empowerment. The journey from "her" to "hercut," from the outermost to the innermost narrative, is not a simple jump. It involves going through the history of women's tradition that occurs in the second narrative fold, "hercut2/journal" and "hercut3/story." If "hercut2" represents the female body or text that is unaware of itself as it emerges out of the dark background of patriarchal discourse, "hercut3" is its reverse image, in which the dark background is illuminated so as to make the female body or text stand out in its own light. To avoid following the dualistic logic of the male discourse by making "hercut3," a polar opposite of "hercut2," an end in itself, the mediating drawing, "chimera," strategically placed at the core of "hercut3" in the tree map (and as a subheading, "hercut3," in the chart view and the outline), intervenes as it introduces the concept of "hybridity" in any rethinking of the text of the body or the body of the text.

Chimera, the fire-breathing she-monster in Greek mythology, possesses the head of a lion, the body of a goat, and the tail of a serpent. Chimera is also the name given to an individual or an organ that is constituted of diverse genetic material, especially at a graft site that marks the joining of tissues from two different genetic sources. The use of "cutting," "grafting," and "joining," recurrent throughout the text, suggests that unlike the continuous dominant male tradition, women's tradition, when seen in the historical context, has been discontinuous, assuming different guises and forms. A woman writer or artist can connect to this tradition only by a deliberate act of "grafting" recollected lives. This recollection is not a search for a direct line of descent but rather an unraveling of a patchwork of connections that have been the fabric of the lives of women, both literary and nonliterary, throughout the centuries. Then, again, a chimera is also a mental construction that does not have a basis in the real world; it is the stuff of dreams and myths. Literary monsters and hybrids might not be real; they can, however, serve as dream symbols for empowering women's lives in the material world. The multiple refractions that emanate from "hercut2," "hercut3," and "chimera" thus set the stage for the unfolding of the story of the rebirth and metamorphosis of the female monster that exists in time and at the same time marks the temporal process of unfolding female creativity.

The fragments under "journal" narrate Mary Shelley's encounter with and parting from the monster that she creates. The journal entries end with remorse on her part for being unable to give a part of herself to the monster so that she might have continued through her. The narrative thread in "journal" thus suggests how difficult it must have been for women in Shelley's time to write as women. A woman could write only through suppressing her voice or disguising it as a male voice, which, however, was inadequate to describe her own experiences.

What is dark and left unsaid in "hercut2/journal" becomes light and achieves voice in the mirror figure "hercut3" and the corresponding narrative, titled "story." This "story" is a continuation of the story of the female monster in Shelley's narrative. The first part of this section is titled "M/S" (which could stand for monster/Shelley, Mary/Shelley, or Me/She)—again, a play on the name that Shelley Jackson shares with Mary Shelley. It includes long excerpts from Mary Shelley's *Frankenstein*—the monster's plea to Frankenstein to create a companion for him, Frankenstein's promise to do so, and then his treacherous decision to destroy the half-finished female monster and disperse her remains in the ocean. The female monster, destroyed in Shelley's text, comes alive in Jackson's text and reflects on her destiny: "I told her to abort me, raze me from her book; I did not want what he wanted. I laughed when my parts lay scattered on the floor, scattered as the bodies from which I had sprung, discontinuous as I myself rejoice to be. I forge my own links, I am building my own monstrous chain, and as time goes on, perhaps it will begin to resemble, rather, a web" (Jackson).

The resurrected female monster encounters Mary Shelley once again, but this time they have a fruitful exchange. A mutual cut and an exchange of words as well as skin takes place before their final parting. The monster experiences this joining with Shelley as a "live scar" that marks a parting, which, at the same time, "commemorates a joining" (Jackson). The female monster's metamorphosis goes through several stages (severance, seafaring, séance, falling apart, and rethinking) involving a journey across the sea in disguise, encountering a woman in male disguise, forgetting her past history, hearing ghost voices, remembering her scars, leaving behind her male-defined "feminine identity," and rethinking nomadic identity. At each stage of the metamorphosis there are doubles, hybrids, and other monsters that are multiple reflections of the female monster.

Marking "phrenology / body of the text" and "hercut4/graveyard" as the third fold, the narrative moves forward and outward. In "phrenology" the exposed head reveals different areas of the brain responsible for different memories or thoughts out of which emerge the coherent "body of the text." Similarly, "hercut4" displays the fragmented body out of which arises the body of the modern female monster as revealed in the textual segments under "graveyard." The left and the right sides of the third narrative fold, which describe the constitution of the body of the modern monster and the modern monster's narrative, set in motion another set of relations, which mark the moments of dispersal and reintegration or deterritorialization and reterritoralization that accompany the constitution of female artistic subjectivity as it continually negotiates difference with the dominant tradition.

In an interesting narrative mutation, the section "graveyard" that appears under "hercut4" in the Storyspace map is included under "hercut3" in the chart view, the tree map, and the outline. This suggests that the assembling of the body of the female monster is common to the second fold, which narrates the rebirth of Shelley's female monster, as well as the third fold, which includes the modern monster's narrative. The common birth brings the two narratives together, turning the modern monster into a symbol of the female subjectivity that emerges out of the fragments of the discontinuous women's tradition.

In "graveyard" the monster names her body parts, attributing each to a different woman from the past. The body parts are thus linked to stories of ordinary women who had no opportunity to become extraordinary as their creative spirits wasted away under the burden of the social conditions that shaped their lives. The body of the modern monster is thus composed of the clear and calm eyes of the unknown village historian Tituba, the lips of laughing Margaret, the tongue of talkative Susannah, the sharp nose of Geneva, the promising ears of Flora, the trunk of dancing Angela, the finger of nameless scholar Livia, the right leg of unwed and not-yet-crazy Jennifer, the left leg of adventure-starved Jane, the heart of Agatha, the strong foot of Bronwyn, and so on. The body of the modern monster also contains the liver of a man named Roderick. The playful insertion of a male story in a string of women's stories is meant to draw the reader's attention to the fact that the focus in Jackson's text is not on creating an essentialist category called "woman" but rather on laying bare the historical shaping of the female subjectivity.

Whereas the body of the modern monster is made primarily out of the fragments of bodies of women from the past, her narrative in "body of the text," occupying both the periphery and the center of the text at the same time, reflects on the multiplicity of the female subjectivity as well as the hybrid nature of all texts. The modern monster describes herself as a double agent who is both whole and dispersed. She incorporates within herself personalities or memories of fictional, real, or disguised women. The women of the past "draw together, bound by a hidden figure that traverses them all" (Jackson). Writing in between the lines of the male discourse, she describes herself as a whole with "haze around the edges." She urges the reader to "come closer, come even closer: if you touch [her], your flesh is mixed with [hers], and if you pull away, you may take some of [her] with you and leave a token behind" (Jackson). Reinscribing the female subjectivity becomes "a matter of redrawing an outline. Snaking through the space between two lives to wrap a line around some third figure" (Jackson). The third figure emerges as the present and the past, the now and then, and the dominant and the suppressed traditions come together in the interaction of the reader and the text. The encounter is marked not by binary opposites becoming subsumed into a unitary synthesis in search of the same masculine model of subjectivity but by the emergence of a new model of subjectivity that exists in difference without conforming to any essentialist descriptions of one or the other.

Jackson achieves fragmentation to open the in-between spaces by using patchwork as the thematic as well as the structural principle of her text. As Deleuze and Guattari (1987) point out, a patchwork represents a smooth space that has no center because "its basic motif ('block') is composed of a single element; the recurrence of this element frees uniquely rhythmic values distinct from the harmonies of embroidery (in particular in 'crazy' patchwork, which fits together pieces of varying size, shape, and color and plays on the *texture* of the fabrics)." It is through its "amorphous collection of juxtaposed pieces that can be joined together in an infinite number of ways [that] we see that patchwork is literally a Riemannian space, or vice versa. . . . The smooth space of patchwork is adequate to demonstrate that 'smooth' space does not mean homogenous, quite the contrary: it is an *amorphous*, nonformal space prefiguring op art" (476–77).

In Jackson's text the textual patchwork creates just this smooth space in which individual textual units achieve significance both in themselves and in relation to one another. The visual and textual units become the blocks of the electronic patchwork, even as the thematic focus itself revolves around the exploration of the patchwork subjectivity. Another refraction of the patchwork connects it to the women's tradition that originated in needle and thread. The meaning of the text emerges from relationships that are amorphous, coming into play and dissolving as readers thread their way through the fragmented textual landscape. Readers of the text have an experience very similar to that of the narrator of the text ("body of the text"), who/which describes herself/itself as "a discontinuous trace, a dotted line" that is "a potential line, an indication of the way out of two dimensions. . . . Because it is a potential line, it folds/unfolds the imagination in one move. . . . A dotted line demonstrates; even what is discontinuous and in pieces can blaze a trail" (Jackson). As one reads Jackson's text, narrative trajectories create a fabric of interrelations that produce a continuous variation of form and meaning in both texts, the body of the text and the text of the (female) body. The creation of gaps and in-between spaces allows interaction that is liberating both for the writer, who must struggle to find her voice as a woman, and for the reader, who must genuinely engage her text.

Without making any claims to originality, Jackson's text is a continuation of Mary Shelley's story, which is every woman's story both literally and metaphorically. Jackson turns Shelley's text inside out by making what is invisible, voiceless, and undefined in the latter into what is visible, voiced, and definable in its indefiniteness. By giving voice to Shelley's monsters she reinscribes her own monstrous artistic self, which finds its continuity with that of her foremother even as it goes beyond her in self-understanding.

The works of both Jackson and Malloy open up the patriarchal cultural text from within, showing how it is comprised of heterogeneous discourses that speak through the gaps and the interstitial spaces. Through recursive narration unfolding as a series of reinscriptions they reveal the complex discursive matrix that has shaped women's subjectivity for centuries. The electronic medium allows them to create a dynamic text that becomes a theater of interaction and enaction, joining the past and the future in the present moment of the reader's experience. The hypertextual aesthetic is thus rooted in active and interactive reading, as is oral storytelling. Multilinear narratives can be regarded as a return to oral storytelling, which Walter Benjamin reminds us "permits that slow piling one on top of the other of thin, transparent layers which constitutes the most appropriate picture of the way in which the perfect narrative is revealed through the layers of a variety of retellings" (93). Benjamin's lamentation about the death of storytelling in the age of information finds its apotheosis in the birth of hypertext narratives, which continue the tradition of oral storytelling.

3   MULTIPLICITY

Database and Interface

HISTORICALLY SPEAKING, narrative has been associated with the novel and the film. With the advent of the new media, a new category of narrative has come into existence that is intricately linked to the database—a collection of items that constitutes the content of the work and exists as binary code in a computer. Unlike the print medium, in which content is the same as the interface, the database produced by the writer for the digital medium needs an interface to make it accessible to the user. In fact, now the same content can be accessed in multiple ways. Thus in electronic literature or artworks a second layer on top of the content has to be created in the form of the interface. The database logic that governs new media works introduces a whole set of new possibilities for conceiving a work.

In a print text like Mark Z. Danielewski's *House of Leaves* (2000), in which a complex collage effect is achieved through the skillful use of footnotes, letters, stories, stories within stories, different fonts, blank pages, and upside-down pages, it is still possible to acquire an overview of the whole work, at least visually. In an electronic text, on the other hand, no such visual mastery is possible because the electronic narrative unfolds in time, as does film, even though it is different from film in that the database cannot be seen but only accessed. In a digital work, Lev Manovich (2001) writes, the database is permanent and real, whereas the narrative is virtual because readers can trace their own path through the narrative. In film, on the other hand, narrative is real and the database is virtual in that film is the final product of the film crew's work with the database of possible shots and scenes. What was in the background in film is foregrounded in new media works, and accordingly the experience of the reader or user is transformed in that hypertext reading can be seen as exploring multiple possibilities of reading through the database of the work. The linkages and their organization can determine what is visually presented on the screen when the user acts on the text. When linking becomes an organizing principle in a work, it dramatically affects the way the narrative unfolds. Both the context and the content morph with the choices that the reader makes as he or she reads the work. Several kinds of discourses using different media can thus be gathered together in a hypermedia work, with various discourse units enhancing the multiplicity that the work projects onto the reader.

Because the print medium has promoted the convention of a transparent interface, the role of the medium in shaping the reading experience has not been taken into account in literary criticism. The artists' books of the 1960s and 1970s that draw the reader's attention to the materiality of the book have occupied the periphery of literary production and as such have not had any substantial impact on how theorists and critics have perceived the role of the medium on the unfolding or exploring of a particular literary work (Hayles 2002). It is perhaps for the same reason that in early hypertexts, discussed in the previous chapter, the verbal textual segments and the links joining them constituted a major aspect of the hypertext writer's artistic strategy. The early hypertext theorists (Landow, Joyce, Bolter, and others) exclusively focused on the linked structure of the electronic text as they theorized about first-generation literary hypertexts. That could have been partly because of the limitations of the electronic medium during that period, which allowed easy access to functionality geared toward manipulation of verbal text through programs like Storyspace or Hypercard. Another reason could have been that the early hypertext theorists saw the electronic text through lenses coated with the conventions and strategies of reading print text. As electronic literature has evolved, paying attention exclusively to the written chunks of text and linking has been seen by some recent critics as very limiting because this leaves out the medium's contribution to the reading experience.

Janet Murray describes various elements of the digital environment in terms of procedural, spatial, participatory, and encyclopedic components. The technical or procedural components have to be dealt with in order to operate in the new medium. The spatial, participatory, and encyclopedic possibilities allow the creation of narratives extending out in space both visually and experientially. In electronic works, no longer do the reader's actions alone determine the course of the narrative. The interface design can also contribute to the meaning-making process, which has dramatic implications for both the writer and the reader. Software applications make it possible to create a work with text, sound, animation, or image, each of which can be programmed to appear onscreen in a variety of ways. Onscreen displays can be made reversible or irreversible through programming as the reader interacts with the text. Various components of electronic narratives—for example, the database of content, the material interface, and the mouse-overs or keyboard clicks of the reader—destabilize the earlier one-to-one relationship of the reader with the print text while at the same time bringing to the forefront the materiality of the medium. To what the medium makes possible with respect to accessing the text is added another dimension of the role the medium itself plays in the meaning-making process. Not only are the reader's eyes and hands engaged in playful interaction with the text, but the work itself can acquire a certain degree of intentionality. The user interaction is thus only one of the elements shaping the screen display.

In complex hypermedia works of literature, there is a dynamic relationship between form and content. Such works retain the best of print literature in their artful use of language, imagery, metaphors, and various literary devices while exploiting the potential of the electronic medium to the fullest. Katherine Hayles (2001) proposes the term "multicourse" to describe hypertext literature that is comprised of various discourses that can be explored by the reader in multiple ways. The hypertextual breaks in the narrative create temporal and spatial dislocations that mark the points of disruption as well as providing frames for alternative narratives whose relationship to the main narrative is parallel and extrinsic or embedded and intrinsic. For example, the hypermedia work "The Ballad of Sand and Harry Soot," coauthored by Stephanie Strickland and Janet Holmes (1999a), brings together a unique poetic text with images of digital artworks originating in a variety of disciplines to create a web of relations with limitless potential for diverging and converging series of readings. In a more recent work, *V: Vniverse* (2002), coauthored by Stephanie Strickland and Cynthia Lawson, the programming of the interface was done in such a manner that the assembling and disassembling of the semantic and graphic elements takes place in a variety of ways. In both works the database elements are broken into their smallest meaning-making components and programmed to be assembled into larger or smaller semantic units through reader interaction with the text. The graphic design of the interface adds further nuances of meaning to the unfolding text. In the hypertext environment, therefore, the visual dimension of seeing the text becomes as important as reading it.[1] The physical interaction of the reader with the text adds other layers of meaning. Using Hayles's media-specific analysis to read Strickland's works shows that paying attention to "materiality allows us to see the dynamic interactivity through which a literary work mobilizes its physical embodiment in conjunction with the verbal signifiers to construct meanings in ways that implicitly construct the user/reader as well" (Hayles 2002, 130–31).

"The Ballad of Sand and Harry Soot"

The print version of "The Ballad of Sand and Harry Soot," written by Strickland, appeared in the *Boston Review* (1999) and was the winner of the Boston Review Prize. Reading the print version is a totally different experience from reading the hypermedia work with the same title. The linear unfolding of the ballad in the print version serves to keep all readers on the same track. The individual reader can still perform imaginative leaps, but only in the realm of Sand and Soot, who emerge as the central characters in the ballad. In the hypermedia work the Sand–Soot center of the narrative is destabilized as links within the ballad and links to images provided by various contributors from a variety of disciplines create multiple reading tracks within the hypermedia work. Other narratives make inroads into the unfolding drama between Sand and Soot, who become amorphous as they seep into images that unleash other narratives.

The web-based "Ballad" was published by the *Word Circuits* website. The gathering force of this hypermedia work is the poetic text, embedded in a rich context of images. The images move into many web spaces filled with algorithmic art, webcam images, scale-inversion experiments, hyperbolic geometry, digital sand etchings, and so on. The migration from print format to the electronic environment decenters the text and sets in motion diverging or converging series of readings that exist only by the return of the others.

Thematically, the "Ballad" is about unrequited love between Sand and Soot; at another level it is about the art of navigating through multiple discourses that constitute human experience. In some ways it also alludes to the computer-generated electronic spaces and the humans who interact with these spaces. The sophisticated conception and design of this hypermedia work bring together a variety of discourses from art, science, mathematics, philosophy, and even mythology to create a web of relations. In spite of the centrality of Strickland's text in this hypermedia work, Strickland is listed as only one of the seventeen contributors to the work. Even the act of transferring the poem to the web environment is unbundled in that the title screen states that the poem text is by Strickland, the design of the hypermedia work was done in collaboration with Janet Holmes, and the implementation to the web was by Janet Holmes. The authorship is thus diffused and distributed at various levels, even as the decentered center of the work continues to be the ballad. The hypermedia format allowed Strickland and Holmes to design the work in such a way that the verses on each screen are linked to other verses in the ballad through three links per screen. Each link is either through the image or through words in the text of the poem. Strickland employed the form of the ballad, the earliest form of literature connected to communal gatherings or dance, to serve as the text, and the work itself has been conceptualized as a dance. Commenting on the creation of the "Ballad" in the essay "Seven Reasons Why Sandsoot Is the Way It Is," she says:

> It is akin to choreographing a dance piece for a theater in the round. Every step, every posture, every gesture, has to work for every seat in the house. In addition, in hypertext, the dance has to work even if members of the audience enter and leave at different times. It is also analogous to visual artwork that sets out to be readable at any scale, as for instance fractal work does. We do hope to leave that more coherent impression one takes away from a dance or scale-crossing image.

The aim of the hypermedia work is thus to present a very integrated piece that brings together different discourses in a seemingly coherent fashion while providing sufficient openings that readers can relate to it from many different perspectives. The narrative coherence is reflected in the selection of images as well as the carefully thought-out links in the text. The navigational paths provided enlighten rather than frustrate because readers can choose a complete reading, a random reading, or a link-based reading. The relationship between the verses and the images is revealing. Strickland's "Ballad" assumes the form of a thread that passes in and out of the images that constitute the nodes of the ballad. Thus there is a very intimate relationship between the images and the text of the poem. It is sometimes hard to distinguish whether the verses were written to weave together the images or the images were used to hold the ballad together as a coherent piece. The images do not represent what is stated in the verses; rather they visually express the forces and relations embodied in the verses. Labeling the images as merely the background or the context of the ballad is not accurate, because there is a continuous interpenetration of the foreground and the background, the text and the image. The verses are cryptic, almost idiomatic, so that the reader has to go back and forth between the verses and the images to arrive at various readings.

In an attempt to make the medium reflect on itself, the binary coding, the technological basis of all electronic productions, has been turned into a literary device to frame the verses on each screen. The verses about Sand appear under 0s and those about Soot are under 1s, though in a few verses the numbering is intermixed to illustrate the interpenetration of two realms encoded by Sand and Soot. Each screen reflects a juxtaposition of two voices and two modes of being and becoming: the male and the female, the silicon-based entities and the carbon-based entities, the world of vectors and dimensions and the world of concrete materiality, the world of dreams and the world of reality, the world of flow and the world of rest, the mother-child pod and mother and child as two separate entities. Certain words or sections of the verses are color-coded to reveal connections between the words or parts of the verses listed under 0s and 1s, though no set formula is used for color-coding the words. The function of the image on each screen is then to point toward a space that goes beyond dualities of all kinds or that marks a space of coexistence or dynamic relationship between what is represented by the 0s and the 1s in the poem.

The images of the artworks, which in their original form appeared as kinetic light sculptures, kinetic sand etchings, quilt wall hangings, webcam art photos, or photographic images, serve to translate discourses from mathematics, science, and art into visual metaphors. The text and image on each individual screen are linked to other textual and visual elements of the ballad in a hypertextual collage. In the instructions given under the heading "how?" the reader is provided with three ways to read the work: reading randomly by clicking zeroes on the navigational bar, reading through linkages, and reading the work completely through clicking images on each screen. Another way to read the hypermedia work is through the "Coda" section, which lists the images. A short statement about the contributors accompanies each image or set of images. Also, a link to the websites of contributors is provided. Clicking the images of the "Coda" section takes the reader back into the poem, whereas the web address given under each contributor statement allows the reader to explore the images in their original context in greater detail.

Reading via the "Coda" section is totally different from reading in the first three ways, which give primacy to the poem. Even though the author(s) state that the "Coda" pieces illuminate the ballad's "theme of the passionate relation between silicon- and carbon-based life," these pieces actually do more than that. The "Coda" pieces bring to the forefront the multiple discourses that shape the poetic sensibility reflected in the poem. The images in the "Coda" section assume the form of a hyperlinked digital quilt sewn together by the verses of the ballad. The "Coda" reading allows the reader to go beyond the literal meaning of the poem to the level that deals with forces, vectors, and dimensions of human experience. The visual presentation of metaphors unique to various discourses of art, poetry, science, and mathematics allows the reader to cross the boundary that separates them from one another. If the reader takes the "Coda" pieces as the starting point of the reading and hops from images to verses and back to images, a shifting and changing narrative emerges that refuses to be congealed into a single perspective or a single meaning; instead it spills out in multiple directions. The center of the hypermedia work is thus nowhere and yet everywhere.

Although a variety of images are used in the "Ballad," those from Ho, a collaboration between Jean-Pierre Hebert and Bruce Shapiro, are central to its conceptualization. The title page of the "Ballad" depicts "Swheel" (Small) from the Western Gallery of the Sisyphus website, tinted by Janet Holmes. "Swheel" is the image of a sand etching, a heptagon with fractal overlapping images of the pattern traced by the Sisyphus ball. The credits page has another image, "Detail of 'Swheel,'" which depicts an enlarged corner of the heptagon with the ball. As the reader mouses over the second screen, the part of the screen that takes her to the poem is the image of the ball of the Sisyphus computer-controlled device used to trace patterns in Ho's kinetic sand etchings. The trackball of the reader's computer and the Sisyphus ball used to trace the sand etchings become one in the clicking; the computer trackball takes the place of the Sisyphus ball as the reader embarks on tracing a pattern in the hypermedia work.

If curious about the Sisyphus ball, the reader can click the "Coda" section and find out more about the Sisyphus project, including the name Sisyphus. The reader can follow the outgoing link to the Greek myth. In the Greek myth, the gods condemned Sisyphus, a mortal who loved life, to the punishment of rolling a stone up a hill for eternity in the underworld. Once he reaches the top of the hill with the stone, it rolls back to the bottom, and down he goes to roll it up again to the summit. Just as the mythical Sisyphus rolls a stone up a hill for eternity, Sisyphus, the computer-controlled kinetic device, rolls a ball in the sand. The myth can be interpreted in a variety of ways, but what is relevant in the present context is more along the lines of the interpretation that is given by Albert Camus—again read by following an outgoing link from the Sisyphus site. Camus is not so much interested in Sisyphus's rolling the rock up to the top of the hill and, in his hard labor, becoming one with the rock. He is fascinated by the pause when Sisyphus reaches the summit and starts to descend downward to get ready to roll the stone up the hill again. This moment of pause, Camus says, is the moment of full awareness or consciousness for Sisyphus and the moment when he overcomes his fate. The digital sand etchings traced by the ball operated by Sisyphus are the realization of this state of pause, an aesthetic pause that takes the beholder beyond preoccupation with the mundane rituals of life to a heightened state of awareness.

Of the thirty-three images included in the hypermedia work, eleven are of Ho's digital sand etchings (from the Sisyphus project), which are woven into the fabric of the ballad. Sand sculptures or etchings, the most ephemeral of all built environments, can be toppled by a mere touch or blown away by a puff of wind and are meant to remind the beholder of the transience of existence. The reader can once again follow the outgoing link to the Sisyphus project from the "Coda" section and, under "Concept," come across the following statement: "Picture a sand garden in the spirit of the Zen gardens of Japan. A digital system conducive to meditation, peace and serenity: where beauty and nature stroll hand in hand through the rhythm of human existence." The spirit of this statement shapes various readings of the ballad.

Interestingly, there is more to this ballad than the use of sand as a medium and a way to reflect on human experience. In Ana Voog's "Goldshow" images, the woman's body itself becomes a medium for artistic expression. The images are from "Anacam," which is Voog's "24/7" video performance project. Images 1, 3, and 7 of "Goldshow" show a hazy picture of a woman, sitting up or lying down, against a granular, sandy background. The images are bathed in gold, yellow, and yellowish-white hues. Clicking the images takes the reader into the verses that are descriptions of Sand. Thus there are contrasting images of Sand, those that are associated with the Ho images, which evoke austerity, simplicity, and tranquility, and those associated with the Ana Voog images of a sensuous woman's body, Ana's body, revealed through the lens of a webcam. The daily rituals of living are turned into aesthetic occasions for viewers who view the live images as they are posted on the Anacam website.

Strickland et al. thus create a poetics of sand by weaving the images that embody fragility, simplicity, beauty, transience, sensuality, and dreaming into the ballad of Sand and Harry Soot.

The first two verses of the "Ballad" read:

0

Sand was a gourd fanatic
and she played

> a glass
> marimba.

1

> Harry Soot loved to listen. ("Ballad")

The image that accompanies the first two verses on the first screen of the ballad is one of the Ho images called "Three Rocks," which is also the logo for the Sisyphus project. The "one rock" of the Sisyphus myth is transformed into "three rocks" of the Sisyphus project. Three rocks in the sand etching are emblematic of the theme of the ballad that centers around the player, the listener, and the song, in other words the reader, the writer, and the work. The ballad is about language, words as well as navigation in art as well as life. In a Japanese sand garden the beholder strategically sits outside the garden in a position of rest so as to resonate with the tranquility and peace that are embodied in the sand garden.

The word "listen" in the preceding verse is linked to the following verse:

0

Twirly languid blue-eyed blue pearls clearly not Sand.
Down on the fourth harmonic she simply singly for a second
stood, so symmetric, second subsequent swiftly sliding side-
riding slamjamming shivering switching—

1

Soot calls it "searching."

("Ballad")

The incomprehensible verse under "0" seems to portray the shifting
and changing nature of Sand. Sand is not "blue-eyed" like Soot, as
we find out in another verse. This verse thus alludes to the ocular
metaphors that dominated western discourse for centuries. The rise
of ocular metaphors in the seventeenth century was closely linked
to the development of the science of optics, which made it possible to see both the large and the small more accurately. But the
sole reliance on the eye also eliminated other perspectives and other
ways of seeing. As the reader searches through the jumble of sliding
and slithering words, the eyes fall on the accompanying image of a
kinetic light sculpture by Friedlander called "Spinning String Light
form." This is what Friedlander says on the contributors' page about
this work from his "Visual Music" series: "In this illustration, you see
a spinning string vibrating in harmony, this description sounds like a
musical instrument, but it is a light sculpture. The vibrating form is a
superposition of the second and fourth harmonic: a 'visual chord.'"
Relying on this statement, Sand's movement and shape-shifting can
be seen as visual music: a transposition of the sense of seeing into
the sense of hearing.

The hypermedia work thus serves to bring the two domains
together: the inner and the outer space, the mind and the body, the
abstract and the concrete, the real and the virtual, science and art.
The crossing over from one discourse to another is illustrated by the
image "Henon Strange," by Brian Meloon. The contributor statement
about him reads: "The Henon map is the two-dimensional analogue

of the logistic equation $H(x,y) = (x^2 - a^*y + c, x)$, where a and c are constants. The picture shows a portion of the locus of points where the rate of escape of a point under iteration is the same going forwards and backwards (using the inverse map)." This statement might not mean much to those who do not rely on numbers to make a living, but the accompanying map is comparable to any work of art. It has a musical quality about it, with colorful flowing shapes that have a rhythm of their own. The image is linked to a screen that describes Sand as a marimba player.

Another verse comments on the multisensory as well as synaesthetic perceptions of Sand; Soot, however, seems to rely primarily on visual perceptions.

0

As albino cave bats who let go of
coloration, but develop keener sensors,
Sand.

1

Soot, who seeks to catch
a falling star in the monitoring
cave, evolves into colorblind. ("Ballad")

This pair of verses is accompanied by the image of one of the Ho pieces called "Winddrifted," which depicts a sand etching in the process of being erased. These two verses contrast two different perspectives. Sand, by giving up attachment to one particular form or shape, embraces many forms, shapes, and perspectives. Soot, on the other hand, with his monocular vision, is oblivious of other colors or perspectives. In an interesting series of links, the word "bat" is linked to "Fibonacci x 3," an image of a patchwork quilt created by Elaine Krajenke Ellison, a mathematics teacher. The biographical section on Ellison states that she "has designed and made numerous quilt wall hangings to inspire her students to explore mathematics from a new perspective." In the original piece "11 different fabrics are used for the design, which is enhanced by the quilting lines that extend outside the triangle." The reader also comes across another quilt image by Ellison called "Poincaré Plane." A precise description of the quilt is given: "Poincaré Plane" is "44 inches in diameter; white background fabric with contrasting bright blue fabric for the tiles, which were sewn on using an appliqué technique." Another quilt image woven into the ballad is that of "Spiraling Pythagorean Triples," designed and quilted by Diana Venters (also a mathematics teacher). Once again, a precise description of the quilt is provided: "37 x 37 inches; 7 different brightly colored fabrics and black fabric. The design center is a one-quarter-inch square; the side lengths of the largest triangle are 3.25 inches, 21 inches, and 25 inches." Quilting has traditionally been a women's craft and a way for women to express themselves artistically. In the old days patchwork quilts were made out of scraps of material cut out of discarded clothes. These quilts were a part of family history. In Ellison's and Venters's quilts the art of quilt-making is used to make mathematics more accessible. Both Ellison's and Venters's quilts bring together the discourses of feminism, art, and mathematics.

The word "evolves" in the preceding verse is linked to the screen with the following two verses and an image titled *Hilbert.*

0

If a silly con were all Sand were.

1

If an ashy trash were all of Soot. ("Ballad")

These verses seem to pose a question regarding the nature of Sand and Soot. Is Sand just "silly con" (a play on the word *silicon*)? Or, in other words, can Sand be reduced to its chemical composition? This verse alludes to Sand both as a metaphor and a medium. The sand is composed of silicon dioxide, commonly called silica, which is used to make silicon chips for computers. Computers are capable of generating electronic spaces of immense simplicity or complexity or art objects of subtle beauty. The online world is a sea of information—a space that requires navigation; it is also a medium of communication, including artistic communication. Silica is also used in the manufacture of glass, the material used for Sand's musical instrument. Should we then judge Sand by her mediumistic potential? Sand's world is that of open possibilities; it involves both creation and navigation. This reading is validated by the verse on navigation that is linked to the term "silly con."

How do we approach Soot's existence? Is Soot simply a lump of flesh that one day will turn into ashes? The word "ashy" in the preceding verse connects to the first screen of the ballad, which presents Sand as a "gourd fanatic" playing a "glass marimba" and Soot as listening to the music. The image "Hilbert" accompanying the preceding verses depicts a ball in the act of creating a sand etching as it rolls through the sand. Sand's playing of a glass marimba and the Sisyphus ball in the act of tracing a sand etching refer to the creative act itself. Once again the linked verse reveals to the reader that, like Sand, Soot should not be reduced to his materiality. The ballad tells the reader of Soot's growing love for Sand and her mediumistic potential to generate diverse worlds of great profundity and subtle beauty. Both Sand and Soot are thus constituted and reconstituted as the reader traverses multiple paths through the hypermedia work. The linked verse on navigation reads:

0

Sand, a cat's cradle fan and economic.
Her shave and a haircut, fifteen cents;
her Oceania nodes of knot
    remembered navigation;
her numerous fingers interlaced
with gloves—made of holes—slipped
successfully over;
    her mediumistic con
in the dark apparatus, all one, all
the same nano-rope. This point
escaped Harry. Harry preferred Ouija
wavering words, reassured by
Ouija jerk. ("Ballad")

Sand in the preceding verse is associated with cat's cradle, a game between two players using a string to create patterns. Just as the same sand, one medium, can be transformed into different shapes, so can a piece of string or rope be used to create multiple patterns or string figures that have become an integral part of many traditional cultures. The mediumistic potential of both the sand and the rope can be used to create new things. The image that accompanies the preceding verse is "Verenga-uka" (Female Spirit), a string figure from Easter Island. Sand is thus intimately associated with her creative potential; in fact she is described in terms of what she creates, which is referred to as her "mediumistic con." Soot, on the other hand, relates to the world primarily through what he sees and what he can describe in words. Soot thus lacks the ability to navigate through the sea of multiple discourses that constitutes human experience. By juxtaposing the Sand and Soot verses on each screen and intermixing the two in a few verses, the poem seems to say that Soot is opening up to Sand's mode of perception. At another level, Sand and Soot can be thought of as a composite and the poem seen in terms of Soot's opening up to the "sandy" parts of himself.

Sand morphs from one shape to another, now a symmetrical or an asymmetrical pattern and now swept out of existence by a slight gust of wind. The transience of sand patterns extends outward to the transience of all creation. Thus the "Ballad" is not only about navigation but also about time. Because time and memory are intricately linked, it is also about memory—in fact, about two types of memory:

1

. . .

Soot is attached to his memory lines,
crow's feet crinkle, scar arroyos, worry
furrows, wry sag, time written in skin,
in bone, in blood. Chemical peels do not
appeal to him. Nor implant chips (wait
until he gets sick!).

    01010011
    Sand's unbelievable memory
    learned, of course,
    not lived. ("Ballad")

The two types of memory are personal memory, which determines Soot's being and becoming in the world, and universal memory as manifested in the history of various discourses. Sand thus embodies the memory of multiple discourses that constitute human experience. The image that accompanies the preceding two verses is "Tread" by Trudy Myrrh Reagan, or Myrrh, whose art deals with ideas in science that somehow reflect back on life. The image seems to be of tire treads that have left their mark.

In one of the verses Sand is associated with emptiness:

0

Sand insinuated herself. ZaumZoom in,
she has gone ahead. ZoomTzim out,
    she is not behind. To hear,
in her gourd, her mallet-fall, a relation
to emptiness, finest gauze, so finely
        woven even the strands
            appear to disappear.

1

Harry Soot believes he is watching.
Harry thinks he is in Times Square.
He is. She is not. ("Ballad")

At a more metaphysical level, as reflected in the Ho images that are strategically placed in the text of the poem, the 0 (associated with Sand) alludes to the emptiness or nothingness of Zen Buddhism, and the number 1 (associated with Harry Soot) to the contemplating self. But the Soot verse shows that he has a hard time letting go, because he continues to see what is inside him as what is outside. Thus he continues believing that he is "watching" or looking for Sand outside. Sand is both in and outside of time, a reality that eludes him.

The hypermedia work also plays with the idea of inverting the traditional understanding of the internal and the external environment by turning what is inside into what is outside and vice versa. The scale inversion refers to the central theme of the ballad, which is the relationship between Sand and Soot. Sand as a character is the poetic realization of Soot's creative potential, for which he is "searching." This reading is supported by Alexander Heilner's microbe images. The contributor statement tells the reader that Heilner's work deals "primarily with the intersection of organic and human-constructed landscapes and environments." Heilner's microbe series, the "Airplane Microbes," "Helicopter Microbes," and "Manhattan Microbes," reimagine airplanes, helicopters, and the island of Manhattan as they might be viewed under the microscope. The microbe images are images of built environments transformed into organic entities that occupy internal spaces. The image "Transmission Helix" uses radio-transmission towers to depict a DNA helix-structured molecule that stores and codes information in living beings. The image is accompanied by the following verses about Sand and Soot:

1
Harry is no fool. Harry Soot is shrewd.
Harry has allergies and moods. Harry
    lies—he can't help it.
Harry has structure—genes and grammar.
Harry is a detective, but he can't find
    an answer. Harry is violent
and violently quiet;

01000011
Sand is sand. ("Ballad")

Whereas Soot is solid and has structure, genes, and moods, Sand is ubiquitous. Here Sand as a character in the ballad becomes one with the sand as a medium; the literal and metaphorical levels coalesce. The sand is everywhere and yet nowhere, and perhaps for that reason the sand (and, by analogy, the online computer-generated electronic space) is defined by its own mediumistic potential rather than by its material composition.

0

Sand religiously stops. And starts the next thing.

1

Bluesy Soot can't conclude.

("Ballad")

The sand realizes itself as a discontinuous succession of creative acts. The discontinuity can also be seen as a continuous renewal, which embodies both novelty and surprise. Sand's existence is not marked by the sterile repetition of the same but rather by the repetition with a difference. The same logic applies to the reader's position with respect to the hypermedia work, which enables multiple readings. Each reading is unique, intricately linked to the reader and the context in which the reading takes place. Soot, however, is associated with time and memory, and he can realize himself only in terms of linear progression. He has yet to unlearn his attachment to his personal history in order to fully experience the present moment in its fullness—a moment that is imbued with multiplicity because it comes into existence only to be displaced by a succession of other similar moments filled with creativity and novelty. Soot's encounter with Sand in the ballad can thus be seen as his opening up to a worldview in which creative discontinuity rather than continuity contributes to the enlightened perspective.

In Strickland's "Ballad," linkages are not merely formal techniques for navigation, nor are they visual elements just for decorative purposes. Instead links create a shifting dynamic between the form and the content. The formal techniques enhance and multiply the thematic import conveyed through the visual and verbal elements. In a well-crafted hypermedia work, each facet of the work reinforces every other facet by entering into a relationship with it, thereby creating complementary or contradictory narrative fields that spill out in multiple directions. There is an ongoing interpenetration of the text and the context, of the foreground and the background. The discontinuity and the perpetual overcoming of the discontinuity thus become guiding principles of the work. The hypermedia work by Strickland et al. illustrates very well how artful conceptualization can result in the work's spilling outward into diverging and converging series of narrative trajectories.

V: WaveSon.nets / Losing L'una / Vniverse

Strickland's "Ballad," the hypertext version of which was coauthored by Janet Holmes, exfoliates outward, connecting to works produced by a variety of contributors, but her *V: Vniverse* (2002), a hypertext work coauthored by Cynthia Lawson, is inward-directed and contained. In this work the authors make full use of multiple functionalities of current software applications, bringing to light in unique ways the effect of a well-designed interface on the meaning-making process. *V*, with a dual existence in print (Strickland's *V: WaveSon. nets / Losing L'una*, 2002b) and the electronic medium, lies at the intersection of multiple discourses of science, technology, philosophy, literature, and art. It also represents a direct confrontation of the print and the electronic text. As a self-reflexive work it comments not only on the changing reading and writing practices but also on the transforming and transformed postmodern world. *V* is ideal for exploring how media specificity contributes to the reading experience and what the paradigm shift from modernity to postmodernity implies for reading, writing, and living.

Because Strickland's *V* is a collection of poems with a dual existence, in print as well as electronic media, Hayles's media-specific analysis can be used as a heuristic tool to see how a rhetorical form, for example, a print text, is transformed when it is instantiated in the electronic medium. The question we ask is this: if a print text is transported from the print medium to the electronic medium, does the changed environment in which the verbal text materializes impact the meaning-making process? In the print medium, *V* is an invertible print book with two beginnings, *Losing L'una* and *WaveSon.nets*, both pointing to the middle of the book, which refers the reader to the web-based section called *Vniverse*. It is as if the print book is cleaved into two halves and out of it emerges the electronic version of *V*. In that *V* has a dual existence, it allows the reader to become aware of how medium-specific possibilities and constraints shape a text. Strickland has not simply transported the contents of *WaveSon.nets* into the electronic medium, but, in collaboration with Cynthia Lawson, has totally reconceived the material of the print text in creating the electronic *Vniverse*. The *WaveSon.nets* of *V* can be seen as the external database of *Vniverse*, but the latter is a work in its own right, which means that database cannot be equated with a new media work; *Vniverse* is *WaveSon.nets* and much more. The artistic strategies reflected in the design of the interface have added interpretive layers to the work that do not exist in the print *WaveSon. nets*. It shows that the visualization of the navigational space of a new media work is as important as the creation of the database of verbal and graphic materials.

V: The Print Text

Either side of the print *V* could actually serve as the front of the book, though the publisher has arbitrarily chosen *WaveSon.nets* as the back so as to provide the publishing information there.[2] After the reader goes through one part, he or she must physically invert the book to start the second part. By making the reader handle the book in specific ways to proceed with the reading, Strickland brings the materiality of the print book to the forefront. Another strategy used to bring the reader's attention to the medium is by breaking the continuous print text into smaller textual units. In *Losing L'una* the poems are divided into triplets by creating number headings and subheadings that break the linear flow of the individually titled poems while at the same time, by means of differential numbering, connecting many of those poems while isolating others. In *WaveSon.nets* one long, continuous poem is broken into forty-seven numbered parts, even though the text of most of the sonnets runs into that of the one following.

Through the strategic use of numbering, the poet seems to suggest to the reader not to take the linear presentation of the print medium for granted. The broken poetic units can be read in a different order or a different sequence. In a poem called "Errand upon Which We Came" the poet gently coaxes the reader to begin anywhere and skip anything because the text is designed for that purpose. A linear way to read the book is not a better way of reading it than the one that involves taking detours. The poet compares the reader who follows a meandering path to a leaping frog who does not know which elements he belongs to as he follows the arc of his flight. The reader is thus advised to not get stuck in the linear progression of the poem but to take chances and hop from one part to another. The foundation on which the work stands includes not only the artistic strategies used to create the work but also the imaginative universe to which the configuration of textual elements alludes. So in the poem the roots or the language or words on which the work stands are not the object of recovery but are the indefinable and ungraspable, seen in terms of relationships as the reader is asked to dig up the roots to see what lies beyond.

The overall structure of the work is guided by the metaphor of coding at both the structural and the thematic levels. A literary work can be seen as a code pointing toward a predetermined reality in which there is a one-to-one relationship between the words and what they signify, or the coded words can be regarded as generative in nature in that simple words or expressions can appear in complex variations leading to different hierarchies of meaning. If *Losing L'una* is regarded as 0 of the binary pair 01, *WaveSon.nets* is 1. And *Losing L'una* is losing the 1, or in other words, it could be returning to nothingness or to the 0 state—dissolution or disappearance. Thematically, *Losing L'una* refers to various strategies of reading and seeing, whereas *WaveSon.nets*, as one large wave of sonnets spread over forty-seven pages of the print book, focuses on multiple discourses that shape experience. Whereas *Losing L'una* mourns the loss of Simone Weil, metaphorically represented by the disappearance of the moon as the dawn approaches, *WaveSon.nets* celebrates the rebirth of the poet, who now has become one with her muse.

Vniverse: The Electronic Text

The title page on both the front/back and the back/front of the print book includes its mirror image in a dark "wedge of the sky," suggesting that the contents of each part are reflected structurally or themati-cally in the electronic *Vniverse* of dark sky and bright stars. The *Losing L'una* section seems to be structurally isomorphic with the electronic *Vniverse* to the extent that the print medium would allow it, whereas the *WaveSon.nets* are thematically isomorphic with it in that the print *Son.nets* of this section appear in different mutations in the electronic *Vniverse.*

Whether or not it is a reflection, or whether or not it is isomor-phic, the web-based *Vniverse* can be regarded as a work in its own right, though a richer reading results if the print components are read alongside the electronic component. If the print version is seen as an external database for the electronic version, that is an excellent way to see how the text mutates with the change of the medium. In the electronic medium the *WaveSon.nets* are transformed, both in how they unfold and in how the navigational process impacts the meaning-making process. The unfolding of sonnets in *Vniverse* is dynamic because it depends on a variety of factors, including how the reader interacts with the electronic database of sonnets and how the computer responds to that interaction. This is not a one-way interaction because there are some facets of the unfolding sonnets that are not under the reader's control.

Vniverse can be seen as a meditation on the relation among com-puter processes, user interaction, and the sonnets. The reader is invited to enter the universe of *Vniverse,* and his or her interaction with the stars and star diagrams of the interface releases complete sonnets or sonnet fragments. The release of the sonnets is intri-cately linked to the diagrams of star constellations that appear and disappear by means of the user interaction. Navigational space in *Vniverse* is not just a transparent window by which to access the work but an integral part of its signifying practices. In the explor-atory navigational space of the interface, the machine processes, reader actions, verbal content, and artistic strategies used in the work construct the reader's subjectivity. Because hypertext involves the reader's active encounter with the text, the reader becomes an integral part of the topological space created by the interaction he or she has with the electronic text. In fact, the hypertext reading experience can be regarded as a sort of body writing; the path the reader traces marks the materialization of the text as well as the reader's nomadic subjectivity.

Perhaps here again we can refer back to what the poet has to say about navigation in electronic environments. In *Losing L'una* a quoted passage comments on how the shift to computerized navigational techniques has changed the aviator's relationship to the skies. In this shift the direct relationship to the universe has been lost. The same is true in the case of electronic sonnets of *V,* in which the reader's direct experience of the database of semiotic signifiers is mediated by the interface that displays the text after it goes through a series of translations from machine code to digital code to natural language displayed on-screen. The appearance and disappearance of diagrams of star constellations, an integral part of accessing the electronic text, add other interpretive layers. One wonders if the dual existence of sonnets in print as well as electronic space reflects the poet's need to retain the direct experience of the text for the reader, even as she uses the electronic medium and its varied navigational functionalities as well as design possibilities to reimagine these sonnets.

In an essay on the creation of *Vniverse,* Strickland and Lawson compare readers to nomadic travelers of the ice ages whose movements on the ground were guided by the patterns of stars in the sky that they invented. In creating the interface of dark sky with stars and the accompanying diagrams of star constellations, the work evokes the ancient practice of using star patterns in the night skies to navigate the oceans or as guides in planting and harvesting crops. The patterns or shapes that people saw when they grouped stars in the night skies in constellations varied from culture to culture, as did the stories or myths that accompanied these constellations. The constellations and stories were thus symbolic in nature and reflected the world of which they were a part. For today's sedentary reader glued to the computer screen, Strickland produces an electronic sky with constellations or diagrams to guide them in the meaning-making process. The various invented star constellations that appear in *Vniverse* are Swimmer, Kokopelli, Broom, Twins, Bull, Fetus, Dragon Fly, Infinity, Goose, and Dipper. The star constellations appear and disappear as the reader moves the cursor across the screen. The diagrams can be stabilized by double-clicking on any of the stars along the path of the constellation. A set of keywords is associated with each constellation, which serves as a clue to the thematic content of a particular constellation. The diagrams give some sort of fixity to the release of sonnets grouped under each constellation. However, in spite of this fixity, there is movement involved in the release of fragmented or complete sonnets through the reader's mouse-overs or keyboard clicks. The reader is challenged to create his or her own sonnets out of the sonnet materials that appear and disappear as he or she interacts with the text.

The star diagrams serve as a navigational aid and work as the electronic version of the table of contents for *WaveSon.nets* that is missing in the print version. Here we see the electronic form in dialogue with the print form. The *Vniverse* sonnets are the sonnets included under *WaveSon.nets* in the print book but are divided into 232 triplets in the electronic space and programmed to be released through reader interaction either as triplets or as complete sonnets. The number of times the stars are clicked determines the version of a sonnet that is released—individual triplets along the constellation path, the complete print-version sonnet, or a complete triplet-version sonnet. The triplets can also be released by typing the sonnet number in the dial at the right-hand top corner of the screen. Interestingly, the numbering of the triplets is different from the numbering of the print sonnets, with the result that, if the reader types 45 in the dial, the triplet that is displayed onscreen is not the triplet from Sonnet 45 but rather that from Sonnet 9 in the print version. This is because the electronic sonnets are divided into 232 triplet units, whereas the fifteen-line print sonnets are only forty-seven in number.

The release of the complete sonnet is interesting in that one of the triplets associated with it appears in color as other parts of the same sonnet are slowly released. The title of the sonnet, an important semantic indicator in the print version, appears toward the end in this display. In the electronic space of *Vniverse*, therefore, the triplet in color serves as the title of the completely released sonnet and becomes semantically important. The electronic version thus undermines both the sequentiality as well as the top-to-bottom reading practices of the print sonnet. If *Losing L'una* (print text) triplets are compared to *Vniverse* triplets, once again there is a great difference in how they can be accessed or experienced in each medium. In *Vniverse* the reading is time-based in that each reading is unique and dependent on a variety of interlinked factors as the reader interacts with the text. Even though the release of triplets associated with each constellation is fixed, how the reader interacts with each star diagram determines the sequence of the release of either triplets or sonnets, so many new versions of sonnets can be formed. *Vniverse* thus foregrounds the materiality of the medium as it adds to the meaning-making process. In the print version, even though the linearity of the print sonnets is broken through numbering, the reader tends to read across the numbered division to maintain the linear flow of the sonnet. The difference in how triplets appear in the print and electronic versions shows that electronic space is infinitely flexible and mutable from both the writer's and the reader's perspective. The electronic medium has its own specificity, which is very different from that of the print medium.

Reading *Vniverse* with Its External Print Database

V in its entirety is a work not only about how to read but also about how to see, in the literal sense of seeing the poem as it unfolds before the reader's eyes in the electronic starlit sky and also reading it by spelling out the words as the triplets are released a letter at a time. The work is also about what lies beyond this act of reading or seeing because it points to the imaginative universe that is reflected in the navigational space of the work. In *Losing L'una* the poet refers to any discourse as a "fabricated lens" through which to see the world, but if this lens is imaginatively handled, the language constituting each discourse is itself seen as a lens that reflects a world of its own.

How does the poet conceive of the reader of *Vniverse*? The reader is not just to notice the existence of different discourses or images and record them diligently; the reader needs to quiet his or her mind and look beyond the stars of *Vniverse* to grasp "the profound correlation" between the concrete and the abstract or "become part of the conversation that physical truths enter into with numbers . . . musical numbers, scores, patterns, algorithms." To see is not simply to grasp the material reality of a particular object or occurrence as it appears in isolation but to grasp the whole context, the web of relations, in which it materializes and to go beyond that to experience the primordial rhythm or force that permeates it,

1.29

a hand-mind that reaches for
its breast, a mouth not
held back,

1.30

by pattern upon pattern giving way to deeper
grasp giving in to rhythm or
vibration or milk. (*Losing L'una* 7)

In order to grasp the relationships, the reader needs to develop strategies of seeing that involve not encountering the object of study head-on but looking at it sideways so that the focus of attention is the fringes or the edges—the edges where one discourse merges into another. Thus the poet says:

1.13

Advice
from an astronomer: avert
your eyes, look away

1.14

to see better,
to avoid,

the blind spot hidden deep

. . .

in order "to enhance / your ability / to see near threshold of what can"

1.17

be seen. For something right
on the edge, try the blink method: first look away
from, then, directly at. What

1.18

appears, when you turn aside, disappears
when you look back. (*Losing L'una* 3–4)

Thus seeing is not mastering the object of one's gaze in order to fit it
into a predetermined map but rather opening oneself up to its multiplicity. The sonnets are not messages for readers or a code for readers
to break; they are meant to open a channel, a passageway for readers
to traverse in order to hear answers to the questions that they pose to
the text. There is no final truth to be conveyed by the poet, because
the poet has not seen it. Both hand and mind need to work together
as the reader moves the cursor in *Vniverse* to experience the web of
relations that connects human beings to nature and to natural cycles
of life and death and the world in which it unfolds.

The print text of *V* serves as the external memory of *Vniverse* for
both the reader and the writer, and the combined work becomes
more of a journey to explore contemporary practices of reading and
writing. The interplay of the print *WaveSon.nets*, as external memory
or database, and the electronic *Vniverse* sonnets brings to the forefront the materiality of the print text and how it differs from the electronic text. A session of reading *Vniverse* is definitely not the same
as a session of reading the print *V.* The print text makes it possible
to comprehend it in its entirety, because the reader can go back and
forth to individual sonnets to see how they fit into the poet's ecology
of ideas, which the reader has now made his or her own.

Strickland 's *V* not only points to the conceptual universe of which it is a part; it is also an enaction of what it is to live and to create in a postmodern world. The creative vision involves not necessarily mastering all discourses or embracing all cultures but rather opening a channel for seeing the web of relations that connect the worlds and worlds within worlds that we inhabit. *V* reflects on its own origins in multiple discourses of science, mathematics, poetry, philosophy, and biography. For the adult poet, each word holds a world of its own even as it is linked to hundreds of others. What appeared as divided and separate to the poet as a child now appears to her as an interconnected web of relations. Thus the word "circuit" or the word "lens" evokes images of the circuits etched on silicon chips by women who sit in sanitized rooms looking though lenses, marking the chips. But whose markings are these? The question here alludes to the story of the exploited women who etch the marks on silicon chips, but the marks are not their own. Multiple discourses are embedded in this simple question, which brings together the political, the scientific, and the technical in one single question. The reader is thus provoked to engage in a dialogue with the text to ask questions or to find answers as he or she disassembles the discourses that are brought together in the work.

A characteristic feature of a postmodern work like Strickland's *V* is the proliferation of ruptures and discontinuities, which are easier to plan and integrate in a web-based work like *Vniverse* than in a print work like *WaveSon.nets*. The disjunctions and jumps from one element to another become the pathways of forging relationships that give the work its coherence. In the impossibility of arriving at a single unified ground or single meaning in a complex shifting world, Strickland's work becomes an assemblage constituted of heterogeneous worlds that touch, collide, or interpenetrate. Multiple worlds are multiple perspectives not of a single unified reality but rather of multiple worlds and of worlds within worlds.

Such complex digital works, through privileging multiplicity and heterogeneity, provide a more inclusive field of conveying the experience of living in a complex world. This requires a shift in perspective from a vertical depth-based reading that focuses on what the work means to a horizontal surface reading that enables the reader to see how the various worlds in the work relate to one another. The meaning-making is thus processual in nature because it traces the movement from one form to another and from one world to another. The relationships so forged provide deeper glimpses into the work, the reader, and the writer. What emerges out of this shift is not only how to read a new media work but rather what it means to live in a world with competing, contradictory, and interpenetrating realities.

Each session of *Vniverse* appears as an oral performance. A performance is contingent on a variety of factors—the performer, the audience, and the setting where it is performed. Similarly, in reading *Vniverse* all these factors come into play as the reader, the machine interface, and the database enter into an intricate dance. The onscreen display that materializes as a result of the interplay among the medium, the content, and reader has emergent qualities because it is time-bound and irreversible, like an oral performance. The reader used to the stability of print text struggles to grasp the electronic sonnets in their totality by creating a memory theater in his or her mind, but is continually frustrated in that attempt. Perhaps that is exactly the point that the poet is making. Reading sonnets in the electronic medium is not about mastering the overall structure of the work and where individual sonnets fit; it is rather about opening oneself up to the onscreen display and experiencing the relationships that it reveals. The electronic version of the work is thus isomorphic with the world and the cosmos itself, and the reader's attitude toward it should be the same—to take one sonnet or sonnet fragment at a time and open oneself up to its reality.

From the works discussed in this chapter it is clear that the task of assembling new media narratives is complex. There need to be intimate connections among the theme of a work, the verbal and visual elements, and the navigational system through which the reader accesses the text. Electronic literature is still in the process of defining itself. That keeps the attention of the majority of writers working in the medium focused more on innovative ways of mixing the media and less on the overall coherence of a piece. Even a cursory browsing of the extensive list of electronic works at various sites shows that there seem to be no common assumptions about the narrative design of a work, the mix of the media, or the meaning of the narrative. In the absence of any common agreement regarding formal conventions for creating, describing, or interpreting electronic literature, practically everything is included under the term *electronic literature.* It is becoming apparent, as some have pointed out, that in order for electronic literature to become mature, the writers will need to pay attention to the overall coherence of their narratives when they mix media and create hyperlinked works.

4 ASSEMBLAGE

Memory and Difference

A TRADITIONAL COLLAGE, including assemblage and montage, is created through combining materials from different sources that exist in fixed relation to one another, whether spatially or temporally. Earlier scholars regarded collage as the best representation of modernist aspirations to achieve aesthetic immediacy, but in recent postmodern reinterpretations by art historians, cubist collage has been seen as a reaction to the modernist desire for aesthetic immediacy in that such works, in fact, create multiple fields of reality that exist in dynamic interrelations with one another.[1] A collage by its very nature can be seen to symbolize the postmodern condition, which privileges multiple perspectives and multiple discourses without framing them in a unified and unifying perspective. Brockelman comments on the two antithetical views of collage: collage aspiring to "presence" or "aesthetic immediacy" and collage that is anti-representational in nature. It is precisely this ambiguous nature of collage or this oscillation between two opposite-meaning contexts, he writes, that makes it ideal for studying the postmodern condition. In a collage, "sense is something to be *made* rather than secured. . . . The experience of both insists that we learn to live without guarantees of meaning (the reality of 'knowing our place') and opens the possibility for a kind of meaningfulness that we ourselves produce though a process of judgment" (37).

Landow (1999) traces the similarities and differences between cubist collage and digital collage. Both make use of juxtaposition to create an assemblage that blurs the distinction between border and ground, but they differ, too, in that digital representational space is dynamic in nature. The assemblage or collage aesthetics allows writers and artists to embed their creativity in the work while resisting giving it any final or unitary meaning. A literary assemblage is created from the juxtaposition of different worlds, social, psychological, or material; the past and the present; different genres or historical periods. Due to the ruptures and breaks that are created through the juxtaposition, the individual elements retain their uniqueness even as they are part of a whole that is open. The ruptures and breaks "are productive, and are reassemblies in and of themselves. Disjunctions, by the very fact that they are disjunctions, are inclusive" (Deleuze and Guattari 1996, 42). The literary assemblage constituting the whole thus exists alongside its individual constituents without subsuming them into any totality and without unifying or totalizing the work. Works that hold together different, sometimes contradictory worlds make such an assemblage effective in conveying experiences of minority cultures that unfold at the pressure points of the social and the political.

Linda Hutcheon (1989) argues that cultural postmodernism has been wrongly charged with a lack of critical awareness because instead of promoting one specific world or worldview it promotes eclecticism regarding the worlds, worldviews, historical periods, representational media, or strategies involved. Many critically aware writers, however, use the multiplicity of their work to challenge the dominant narrative as they bring together the verbal and the visual, the past and the present, the seen and the unseen, so that what is visible and known can be reseen and reevaluated. Various elements in their work create configurations that do not mean much in themselves but, upon entering into relationship with one another, produce effects both within and outside the work. The work, in Deleuzian terms, becomes a "literary machine" that the reader learns to use not to discover some predetermined meaning but to see how the machine functions in order to create a plurality of meanings. New media writers use the flexibility of the electronic medium to create narratives that spread out both spatially and temporally into converging and diverging narrative trajectories that have dramatic implications for the reading process. Collage aesthetics is thus rooted in active and interactive reading as is oral storytelling; it is based on process rather than product, discontinuity (collage or photomontage) rather than continuity, and interactivity rather than passivity (Lanham; Mitchell).

I use two works by M. D. Coverley (aka Marjorie Luesebrink), *Califia* (2001) and *Egypt: The Book of Going Forth by Day* (2006), to reflect on the "assemblage" quality of new media writing. In *Califia* Coverley mixes music, artwork, fiction, history, myths, and legends as well as photographs and maps to bring to light the buried fragments of the Chumash Indian lives as reflected in their petroglyphs, cave paintings, sacred figures, and designs. The visual fragments add to the already rich symbolism of the verbal elements. *Califia* is a richly layered and complex text in which the official history of the gold-prospecting period in California is intermixed with the buried histories of Chumash Indians (the original inhabitants of California), the exploitation of Mission Indians, the hidden stories of different crime syndicates in the building of the city of Los Angeles, and the unacknowledged histories of the women of two clans over five generations, who keep the myths and legends of their Indian ancestors alive by passing them on to their children. In the skein of dreams of the elusive gold treasure is embroidered the dream of the Indian star lore and sacred knowledge that the women of the Summerlands and Beveridges pass on from generation to generation. *Califia* writes itself, so to speak, as the characters reinscribe and reenact the past in their search for the elusive gold treasure.

Spider Woman in *Califia,* like the mythical Spider Woman of American Indian lore, is the Chumash Indian ancestor who codes the wisdom and rituals of her people in a hand-embroidered blue blanket. In weaving the designs of star constellations and string figures, she weaves the locations of sacred Chumash caves as well as the rituals performed there, symbolized by the string figures. The star maps coincide with the geographical map of the locations of the sacred Indian caves, presumably the sites of gold mines. Unraveling the stellar design becomes the ritual enactment of the spiritual journey to the sacred caves, which is at the same time a material journey to find the treasure of the Califia. The present is unraveled in all its possibilities with the deciphering of the past, and the storytelling in the narrative becomes the act of re-membering the present with the past.

In *Califia* narrative threads mirror one another in their difference as the itineraries traced take precedence over reaching a destination, even as the need for the final destination is continually reiterated. The empty hole that the questing narrators find at the original burial site of the Califia treasure represents the no-thingness at the heart of the text; the quest for the Califia treasure becomes a quest for the text that would displace this no-thingness and lead to re-membering. The blue blanket that holds the key to the possible new location of the treasure is the encoded text that displaces the emptiness of the first reading. The empty authorial space as well as geographical space (the original burial site) mediate between the past and the present and become the location of reinscription, which is essentially performative in nature. Reading Spider Woman's text imprinted on the blanket leads to a re-collection of the oral narrative (cultural history) as well as the women's narrative, which originated in needle and thread rather than pen and paper.

The textual displacement that mirrors cultural displacement is repeated at another level in the framing of the text. The fictitious narrator M. D. Coverley is an empty, genderless sign that mirrors on the outside the gendered author of the electronic text and on the inside Spider Woman, the author of the embroidered text. Even as absence marks the outermost frame of the narrative, it is mediated by the voices of two female narrators, Kaye and Augusta, which lead to the innermost frame of Spider Woman's voice, which is frozen in time and, literally, in space until it is unearthed and read anew. Kaye, guided by the moon and stars and with the ability to use intuition and imagination, is instrumental in reinterpreting the message imprinted on the blue blanket. In the process she awakens Augusta to the history and wisdom of her ancestors. The third narrative is that of Calvin, whose function is to transfer the Califia material to a computer and discover ways to link various events and characters. At one level, Calvin's narrative is the disembodied narrative of the written discourse as he transfers the written evidence to the computer disk. At another level, especially in the Calvin and Kaye docudramas, it is the mediating voice that emerges from the space between the written and the oral discourse as the two strands, material–spiritual or rational–intuitive, come together. *Califia* thus moves both inward and outward, reaching inward to come in touch with its past in oral narrative and in its movement outward attempting to rewrite the oral narrative along with the written narrative that marks the present. The fluidity of the electronic medium allows Coverley to combine word, image, and sound to create an interface that is as much a guide as a means to access the database of diverse narrative elements, ranging from music, artwork, fiction, history, myths, and legends to photographs and maps.

In her recent work *Egypt: The Book of Going Forth by Day* (2006) Coverley made a more innovative use of new software applications to design a navigational system that embodies the thematic content of the text. She created a work that is both visually fascinating and technically innovative as it presents the shifting and interpenetrating relationships between the temporal and the mythical, the personal and the cosmological. To create an integrated navigational system Coverley modeled the interface after Egyptian hieroglyphic writing, which is a complex system involving three levels of communication: figurative, phonetic, and symbolic. The figurative appears in the form of pictographs. The phonemic sound of each pictograph or glyph is determined by the other pictographs constituting the hieroglyphic word. The temporal quality of the sound is thus visualized through its spatial arrangement. In hieroglyphic writing, word and image are intricately linked, as are word and art, and it is in this intricate linking that the symbolic function of the language emerges. The hieroglyphic inscriptions found on temple walls and stone monuments coded the worldview of the ancient Egyptians.

Whereas the philosophy underlying the Egyptian hieroglyphic writing is embodied in the navigational system of Coverley's work, the Isis and Osiris myth of dismemberment and reassembling of the ancient Egyptian god Osiris is the ur-text on which the plot is based. The narrative has three layers realized through three voices that reflect three levels of awareness of the main narrator Jeanette: the first-person account of Jeanette's journey down the Nile, her reflections on the journey as she recounts in e-mails to her sister Nancy, and selections from the *Egyptian Book of the Dead* marking the journey of Isis in the underworld in search of Osiris. The three voices also represent three aspects of human consciousness according to the ancient Egyptian worldview: the *ka,* the *ba,* and the *akh,* which can be loosely interpreted as the animating principle in humans, the individual soul, and the eternal soul. The work can also be seen in terms of bringing together three different perspectives in the same unified screenic space, screen after screen: the objective perspective, the subjective perspective, and the cosmological perspective. The three perspectives are tightly linked in that the plot of the contemporary narrative has echoes of the ancient myth.

Egypt required that a navigational system be devised that joins not only the present with the past but also the static and the fixed with the moving and the changing. Coverley encoded these two types of time in the same inscription system by designing the interface in such a way that some elements appear animated and others fixed. The kinetic effects in the form of changing backgrounds and transition effects are created through flash animations, programming, and video clips that emphasize movement. In addition, the design of the navigational system allows the reader to read the verbal text in any direction, from top to bottom, bottom to top, left to right, or right to left, just as the work itself can be accessed in multiple ways, reading the individual narrative threads separately or in different combinations. This aspect of the navigational system also reflects the theme of the eternal return.

Coverley relied on the hieroglyphic tradition to create a set of registers, with each register assigned a different function. The vertical register appears as an animated pillar that, when clicked, materializes inscriptions from the *Egyptian Book of the Dead.* Jeanette's letters to her sister appear in another animated horizontal panel with pictures of landscapes or objects. The fixed horizontal register underneath, devoid of animation, provides the objective description of Jeanette's journey down the Nile. The vertical and horizontal registers depict two types of time: temporal existence, subject to death and decay, and cosmological time as manifested in the cycles of death and rebirth. The animated panels (the vertical and top horizontal panels) get into a relationship with the static and fixed register that gives the description of the actual journey.

Egypt contains various types of links that provide the reader access to different aspects of the work. It is possible to read the individual narratives separately, but the interlinked plot elements promote a reading in which each screen becomes a collage of diverse elements to be explored in their multiplicity. The reader can also access the background information found in glossaries, maps, and timelines, along with information about the Egyptian hieroglyphs. A comprehensive understanding of *Egypt* emerges only when all levels of the work are seen in relation to one another in an expanding circle of meaning-making process. The ritualistic journey in the underworld, another journey down the Nile, and yet another in Jeanette's mind coalesce in the reader's journey through the hieroglyphic space of *Egypt*.

Coverley's *Egypt* represents a differentiated multiplicity that refuses to crystallize into a single homogenous perspective. Readers interact with the text as they bring their own expectations and experiences to an exploration of how various forces have condensed into the parts of the literary assemblage and what kinds of effects they produce in words, images, actions, objects, as well as characters. As different parts of the narrative are thrown into relationship with one another, the present is reseen and reevaluated with and against the past.

New software applications have given hypertext writers like Coverley tools to literally create a verbal–visual language encoded in the navigational system, where visual elements are as important as the verbal elements in the meaning-making process. In spite of innovative works that have been created in electronic literature, especially hypertext literature, it is still on the margins of mainstream literary production and almost invisible to culturally marginalized communities. Writers who work in the electronic medium must overcome tremendous technical hurdles ranging from keeping their equipment upgraded to learning new software applications and the new skills and abilities necessary for web production. Many innovations in electronic literature so far have been made by early adopters of technology who have devoted enormous amounts of time to breaking into totally new ways of expressing themselves. Much work still needs to be done in expanding the intellectual, social, and geographical boundaries of discussions surrounding electronic literature so writers from diverse cultural communities can participate in this new form of writing and it can become better established in literary communities globally.

5 TECHNOCRACY

Imagined Futures and "Reality"

THE ERA OF CYBERSPACE, virtual reality, software agents, and tele-robotics is creating cutting-edge technologies that are transforming all aspects of life. The digitalization of data that can be instantly transmitted over the electronic networks has extended the human senses beyond what has been possible through just television and radio. Webcams have actually become optical prostheses through which scenes in distant places can be seen on a continual basis as the images are transmitted to an Internet site to be viewed by any-body. The unitary eye of the television screen peering into living rooms has thus been turned into the multiple eyes of the viewers of computer screens looking outward into vast virtual spaces, crafting their own scenes. Computer screens have indeed become doors to invisible and intangible virtual spaces that materialize for the users only in the moment of interaction.

Promoters of cyberspace set up a dichotomy between the virtual and the material world, describing it in terms of literally emptying the mind from the body into invisible virtual spaces. As Nicholas Negroponte writes in *Being Digital,* the conversion of atoms into bits might be increasing, but we still need atoms to interact with the bits. If being digital is a state of being hooked to the network, with one's point of view flying through empty space, it is being material that makes that connection a reality. Human experience is located in time and is dependent on the body for its actualization in both the real and the virtual realms. Disembodied information can travel across the globe on electronic networks; embodied humans, however, cannot.

Nowhere are humans more different from intelligent machines than in their embodied status and the situated nature of their experience. The construction of the "disembodied" human in the dominant technorationalist narrative makes the need or, for that matter, the desire for a politics of social change superfluous.

The same applies to the construction of race and gender on the Internet. The experiences of cyberspace were initially described as free from the markings of race and gender. Stone and Nakamura, however, show that gender and race markings are very much present in cyberspace. A visit to any online social environment, whether an online role-playing game site or a chat room, reveals that the user or player might be free to adopt a persona, but the nature of the persona is determined by regulatory gender norms. If cyberspace is seen as an extension of the material world, the performance of material bodies in virtual spaces cannot suddenly be conceived of as free of markings of race and gender. These specificities are not simply masks that the users take off as they traverse cyberspace; instead they constitute the very materiality of the users' existence and hence color their thinking and their imagination. Not only are role-playing cybernauts operating from their bodies; the personas they assume conform to the regulatory race and gender norms. Even though cross-dressing is common, the marks of stereotypical gender distinctions are nonetheless preserved.

Humanized Neural Nets and Cybernetic Humans

With the ushering in of the computer age, the machine metaphor of earlier centuries is transformed into the computer metaphor. Although computational models of mind dominated earlier cognitive science as well as artificial intelligence research, in recent years there has been a shift to a "connectionist" or "emergent" paradigm based on the brain model that describes the mind from the bottom up at the level of interconnected subnetworks that stabilize into global behavior. Instead of looking at the mind as a rule-governed mechanism, connectionists conceive of it as a system of simple unintelligent components or interconnected subnetworks that work together in producing global behavior. In the absence of a central processor or abstract symbolic representation at the top level, the system works at the subsymbolic level, from the bottom up; it is thus rule-described from the bottom up at the implementation level rather than rule-governed from the top at the level of representation. For example, as Marvin Minsky, one of the founders of artificial intelligence, writes in *The Society of Mind*, the mind can be seen as a society of autonomous agents. Both the traditional cognitivist model, based on the manipulation of symbolic representations, and the emergent or connectionist model, ultimately describe the mind as an information processor. Whereas the cognitivists describe cognition as grounded in the representation of a pre-given world and pre-given subject, the connectionists go to the other extreme and reduce the mind to a decentralized postmodern machine (Dreyfus; Dennett; Franklin).

Some fiction writers have explored the limitations of such reductive theories through bringing out the complexity of what it means to be a human. For example, in *Galatea 2.2* Richard Powers casts a writer's mind opposite an artificially created mind or a neural network, thus bringing to focus the cognitive, affective, and imaginative dimensions of the human mind. In Powers's text the role of the human as the center of any meaning-making process is highlighted in the structure of the text, its organization, and the multiple reflections of characters, as well as in intertextual references. The fictional protagonist, also named Richard Powers, joins the Institute of Advanced Sciences at a midwestern university as a visiting humanist. There he agrees to assist Philip Lentz, a cognitive scientist, in training a neural net to mimic the text-explicating ability of a twenty-two-year-old English major. In the encounter between the cognitive scientist Lentz and the humanist Richard are embodied two different views of the human mind. For Lentz the mind is simply an associative matrix consisting of multidimensional vectors of stimulus–response shaped by complicated feedback and massively parallel pattern matching. Because Lentz seems to believe that the mind takes shape out of the disembodied fragments of the cultural script, it is natural for him to regard the socialization of the simulated human Helen as possible simply by feeding hundreds of texts into her memory.

Richard reads literary texts to Helen and wonders what these stories mean to a neural net. Did Helen do more with them than simply extracting different associative patterns? From her occasional responses it is obvious she has no real understanding of what she says. She has no concept of self, past or future, and no concept of space. She has no qualitative subjective states of awareness. In fact, she has no intrinsic mental state because she lacks self-awareness. Unlike Helen's memories, which are always object-centered, Richard's memories are integrally linked to his subjective experiences of love, joy, sadness, and loss. The repeated incursions into Richard's mind are used in the narrative to highlight the subjective states of the human mind in spite of decentered brain processes. Even as Richard realizes the hollowness at the center of Helen's being, he nonetheless becomes emotionally involved in his project and starts seeing Helen as a psychological machine that is able to feel and think, when in reality it only seems to do so. In the end, Richard's interaction with Helen turns out to be very similar to the classic psychotherapy session with the computer program ELIZA, which responds intelligently to the user's questions or statements because it is programmed to match its responses to the user's statements.

Galatea 2.2 reflects that researchers understandably must abstract from the complexity of the human mind a simple picture that becomes the starting point of the construction, invention, and discovery of new and complex theories or artifacts. However, this simplified and abstract picture then becomes the basis of constructions of the mind and the body and the relationship between the two that ignore the embodied subjectivity of humans as well as the mediating role of techniques that create hybrid subjectivity and agency. Following the modernist dream of scientific objectivity, the researchers relegate human subjectivity to the background in their enthusiasm for objectifying the mind both metaphorically and literally in the artifact.

The philosophical implications of the connectionist or emergent model in artificial intelligence research have led to mechanistic theories that reduce consciousness to an epiphenomenon of brain processes and the mind to an information processor. Katherine Hayles

(1997) argues that when humans are reduced to information processes the result is underplaying or erasing embodiment, which is an integral part of the human condition. She attributes the trend toward disembodied constructions of the posthuman to the tension between inscription and incorporation or information patterns and material embodiment in contemporary information culture. In inscription technologies based on coding, the signifiers achieve significance outside of their materiality because they can be transmitted from one place to another free of materiality. When inscription is privileged over incorporation, it is easy to lose sight of the embodied nature of humans as well as the communities of which they are a part at both the local and the global levels. Hayles believes that this promotes the view that information contained in the DNA or transmitted over data networks somehow exists separately from the context from which it arises. In cybernetic models of subjectivity, therefore, "embodiment in a biological substrate is seen as an accident of history rather than an inevitability of life," consciousness is turned into an epiphenomenon, and the body is seen as an original prosthesis that humans have learned to use; accordingly, computer simulations or artifacts become just an extension of human prostheses that disembodied consciousness is born with (242).[1]

As the use of biological and social metaphors becomes popular in describing intelligent machines, another configuration of ideas is taking shape about humans plugged into technological networks. In many literary narratives dealing with cyber scenarios, technology is presented as an invasive force that becomes the unexplored frontier where both the human subject and the body are reconfigured and redesigned, leading to transcendental visions of other modes of being or earthly visions of human surrender to technological innovations. As the interior space extends outward through the mediation of technology it becomes a fertile ground for the manipulation of human desires in the infinite cycle of production and consumption. According to Hayles (1999), such cybernetic constructions reconfigure the human "so that it can be seamlessly articulated with intelligent machines. In the posthuman, there are no essential differences or absolute demarcations between bodily existence and computer simulation, cybernetic mechanism and biological organism, robot teleology and human goals" (3).

The construction of the posthuman subject is generally in terms of dispersal, fragmentation, and alienation, as seen in most of the cyberpunk genre of the 1980s and 1990s. Haraway's liberating cyborg myth is turned into the myth of enslavement, in which human–machine coupling is not performed to achieve some liberating political goal but instead becomes attractive for its exoticism or shock value in a culture that is perpetually in need of new fixes. Cyborg technologies thus become technologies of exoticism, which, like cyberspace itself, are avenues for further circulation of capital when traditional markets become saturated. To serve the cause of difference in a consumer society in which there is a competing force aimed at keeping everything the same can itself serve the interests of corporations by promoting the flow of capital.

Media play a great role in shaping and promoting images that have a potential to enter into the network of production and consumption. Recent reality shows on television, such as *Extreme Makeover, Swan,* and *American Idol,* are great examples of how media images of wellness, beauty, talent, and confidence are circulated among viewers as they themselves enter into the production and promotion of images. These images create a mass desire to have a certain kind of body and facial features, which are to be achieved through cosmetic surgery. However, this need for perfection is not confined to procedures to alter the surface of the body. As advances are made in biotechnology, medical interventions are sought at the genetic level.

Nobody can deny the potential of identifying, manipulating, or even eliminating disease-causing genes in relieving the suffering of countless people. It is also true that biotechnology holds great promise for creating potent drugs to treat illnesses or even for creating organs and tissues for patients in need of organ transplants. However, in a culture that strives for perfection, any variation from the norm can be subject to genetic manipulation. Some of these developments are disturbing because they make us think for the first time that technology has the potential to implicate the most intimate parts of the human body, for example, the genetic blueprint that makes us who we are. It is one thing to tamper with the genetic blueprint when the results are confined to the individual but quite another to manipulate genetic material at the level of germ cells, because such changes could then be transmitted to that individual's offspring, thereby altering the genetic makeup of future generations. And no one knows the long-term effects of artificially created mutations.

Earlier biologists looked at the mutation of genes in a positive manner in that variations left open multiple possibilities for the species to respond to the changing environment. Today genetic variations are seen by microbiologists as "errors" in the code. So in theory, editing the errors should make it possible to reprogram the organism. Parents who can afford genetic therapy can request prenatal procedures to introduce desirable traits into the genetic blueprints of their children in order to have perfect babies. Because what is perfect and normal is culturally determined, who is to decide where to draw the line with regard to such risky procedures, which could impact the long-term future of humanity? Biotechnology tools, when fully developed, can be used for purposes that are guided by market considerations more than by the long-term benefits of humans and the environment. Moreover, focusing exclusively on the genes to find answers to all problems, physical, psychological, and environmental, can have a dramatic impact on how people see wellness and illness as well as what is normal and what is abnormal. Corporations that fund big science are not interested in the long-term impact of biotechnology on humans and the environment. As Rifkin rightly comments, it is more profitable to reduce complex physical, social, and even psychological problems to malfunctioning genes, which takes attention away from the malfunctioning of social institutions, the family, the government, and the health care industry and fills the coffers of the multibillion-dollar pharmaceutical industry and the emerging biotechnology industry. In contemporary medicine the focus is on illness-based medicine rather than preventive medicine. Advertising and the mass media are pressed into service to promote these ideas so that people themselves will start believing that this is the only path that science and its technological applications can take.

The innovations in technology cannot be seen apart from the socioeconomic context in which they unfold. In a capitalist society technical rationality uses knowledge in the form of science and technology to dominate and control people. Take the example of developments in the field of nanotechnology. Nanotechnology is a material science of building biological machines of molecular size that subvert the distinction between what is real and what is artificial. Many different technical configurations are possible with nanotechnology, but what use is promoted depends on who funds the research. It is reductionism at work when experiments creating and disseminating nanomaterials are funded without setting aside adequate resources for risk assessment to see how the introduction of such particles would impact the biosphere. It is possible for nanomaterials to have disastrous long-term consequences for human species as well as the biosphere. The same applies to working with plant materials and creating genetically engineered organisms or plants to serve specific purposes. It is not really known what the long-term impact of such genetically altered species would be once they were released into the open and interacted with naturally evolved species. This means that there is less emphasis, or almost none, on keeping the whole picture in mind.

As new technologies come into existence with heavy investment in research by large corporations, the designs that are promoted are not necessarily in the best interest of people or the environment. The new media technologies have made it possible for those in power, whose thinking is primarily guided by market considerations, to reach the majority of people in multiple ways and to influence their thinking. George Orwell said that whereas in dictatorships people are controlled through the use of explicit force or regulations of censorship, in free societies more subtle and sophisticated methods of control are used so people voluntarily submit to their own subjugation. The advertisement-filled television shows and movies of our day, and even the Internet, promote the interests of a minority who are at the top of the economic chain. The attention of the disempowered majority is kept focused on the need to become perfect consumers in order to be happy. Instead of providing an avenue for free discussion and reflection, the wheels of the media work in ways that contribute to people's disempowerment because they become active participants in the simulated world created by the media.

Technocracy and "Reality"

Theorists of technology have commented on how technology has become embedded in its own ideological bias, which means that inherent in it is a set of beliefs that are invisible to its users. In *Technopoly: The Surrender of Culture to Technology,* Neil Postman traces the history of the development of medical technology to show how the introduction of medical instruments has resulted in changes in how doctors perceive themselves and how they interact with their patients. The increasing use of medical technology has also given rise to experts and specialists who know the patients only through their X-rays and other diagnostic tests. Because people unquestioningly accept modern technologies, they stay unaware of the hidden biases of these technologies. And because they are unaware of the hidden assumptions and biases, they let themselves be ruled by them.

Technology is central to modern society, and people's choices about how they choose to live cannot be easily separated from the capitalist technological networks in which they are integrated. In *Transforming Technology: A Critical Theory Revisited,* Andrew Feenberg writes that he accepts that technology is not neutral, but it also is not an autonomous force with a logic of its own. So refraining from using technology by living a simpler life or seeking refuge in religion or nature for spiritual renewal is not the proper way to address the issue of technology. It is not technology that determines a civilizational future but human actions. Current theories of technology, Feenberg further argues, propose that technological development can lead to only one type of society, a society patterned after western capitalist democracies. That means that, whether it is in India or China, a society's future will look very much like that of the advanced western societies. Such theories are based on determinist assumptions that grant technology a power and autonomy that it does not have. Feenberg believes that the ills in society exist not because of technology per se but because of the antidemocratic design of technology, which embodies the interests of the capitalists. In a technocratic capitalist society, technical designs serve the interest of capital, and accordingly those designs are put to use on a mass scale that has a potential for efficiency and profits without paying any attention to how they affect people and the environment—whether they subjugate or empower people, affirm their cultural values, or destroy them.

The dynamics of importing western technology into the developing countries is the subject of Neal Stephenson's *Diamond Age*, which is set in twenty-first-century China. The text presents the role of highly commercialized media images transmitted over mediatron (television) screens, the use of interactive books in teaching young people ethical behavior through universal archetypes of different cultural traditions, and the potential as well as the hazards of nanotechnology. Underlying all these themes is the basic idea that a society needs a moral and ethical framework to function properly. The import of western technology as well as the operational configurations required to implement it has led China on a developmental path that is very much driven by consumer capitalism. The Seed technology, an advanced nanotechnological application, is shown to have the potential to provide an alternative path to the current efficiency-based and profit-oriented western technological model. The aim of the Seed technology is to duplicate nature rather than subjugate it, so it empowers larger numbers of people to be actively involved in determining their lives. If such technology is to be introduced, individuals as well as the culture of which they are a part must develop moral and ethical fortitude to ensure that it is not misused for destructive purposes.

In *The Diamond Age* nation-states have disappeared and society is divided into tribes based on class as well as cultural values instead of race, geographical location, or color. The highly westernized Coastal Republic of China is filled with all kinds of tribes with their own pieces of real estate. The major tribes are Atlantis, Nippon, and Distributed Republic, and there are numerous minor tribes based on common religion or some other shared belief system. Toward the end of the text, even the Hans, the ethnic Chinese, are described as constituting a major tribe because they occupy the largest percentage of land in China and are descendants of an ancient civilization. The Atlantis, the most privileged, primarily Anglo-Saxon tribe, has adopted the Victorian value system based on social and moral codes meaningful to them. The Celestial Kingdom, on the other hand, has adopted the Confucian worldview to run its tribe. It is not advanced in nanotechnology as are the Atlantans, Nipponese, and Hindustani tribes. Dr. X, a scholar and high official in the Celestial Kingdom, subversively acquires the technology developed by major tribes and refines it. The Celestial Kingdom emerges triumphant in the end, defeating the foreigners as well as the Chinese representing western interests who control the Coastal Republic of China.

The text has many interlinked themes, but the central theme is that of educating a young girl using an interactive device, the Illustrated Primer. It is based on the idea that the collective unconscious of humankind possesses universal archetypes that appear in different forms in different cultures, for example, the Trickster figure who appears in different guises in various cultural traditions and is an instrument of change. In the Primer the Trickster figure appears as a Technologist who has control over what is created in the virtual world of the Primer and, indirectly, over what happens in the real world as well. The archetypal figures programmed in the database of the Primer become part of the simulations that guide the thoughts and behavior of users who interact with them. The wise one in *The Diamond Age* is thus the Primer, but it is not wise to think of it as an isolated thing because its effectiveness lies in its existence as part of a cognitive system that includes the user, the narrator, the database, and the programmer of the database.

The Primer is designed by John Hackworth, a nanoengineer, and commissioned by a neo-Victorian, Lord Finkle-McGraw, for his four-year-old granddaughter, Elizabeth. Finkle-McGraw is not satisfied with the education provided in schools. He wants to put technology into service to create an educational tool that will teach his granddaughter to challenge their inherited belief system rather than merely conform to it. In order to create leaders in his tribe, he believes, the younger generation must adopt the worldview of neo-Victorians in a deliberate manner as they themselves find out that this is the best system to use to run an ordered society. Thus he wants young people to learn subversively by interacting with other cultural traditions and in the process learn to appreciate what the neo-Victorian tribe has to offer them in moral and cultural terms. The Primer is soon ready, and Hackworth cannot stop himself from creating a counterfeit copy for his daughter, Fiona, which through a series of unexpected events falls into the hands of a poor girl, Nell, from the Leased Territories. As payment for committing the crime of infringing on the intellectual property rights of the Primer's owner, John Hackworth agrees to Dr. X's request to create mass copies of the Primer with automatic voice technology and other minor changes so it is suitable for educating a quarter of a million Han baby girls rescued from infanticide and starvation in the interior territories of China and now housed on Dr. X's ships. The original Primer required a human narrator, where as the Chinese version of the Primer makes use of automatic voice generation technology.

The Primer hears and sees everything in its vicinity and maps the data in its memory on the psychological terrain of the user. As Nell opens it, it imprints her face and voice into its memory and uses her point of view while creating a psychological terrain for her virtual adventures. Nell's interaction with the Primer results in a detailed branched narrative populated with universalized figures from myths and folklore as well as Nell's life. Stories generated by the Primer involve Nell in interconnected adventures in which she virtually plays out various aspects of her personality as Princess Nell. The narrative grows as Nell's understanding of her world grows. Nell's interaction with the Primer can be compared to the interactions of players of the role-playing games in current online multiuser gaming environments, in which the virtualscape is literally written into existence, engulfing players as they engage in online exchanges with other role-playing cybernauts. The interactive Primer, however, has more functionalities. Nell's virtual friends are not real people but rather simulations generated by the Primer to allow her to develop aspects of herself. Nell's virtual adventures become a practice site for perfecting her skills and various aspects of her personality that will be indispensable to her to succeed in the quest she pursues in the virtual realm of the Primer, the collection of twelve keys so she can set free her virtual brother, locked away in a dark castle. This virtual scene reflects the reality of her brother, a gang member, who is trapped in the miserable life of the Leased Territories, with no hope of release from it. The real and the virtual are thus intertwined in the text.

The success of the Primer in educating Nell is partially attributed to the emotional engagement exhibited by Miranda, a former governess turned actress, who bonds with Nell as she narrates the events in the Primer from a remote location. Miranda chooses to devote increasing amounts of time to Nell's ractive (interactive drama) because she instinctively knows there is a little girl at the other end of the machine interface who needs to be guided. Her guidance comes not in the form of any explicit instructions regarding the choices that Nell's character makes in the virtual journeys that she undertakes but rather in the emotional inflexions of her voice as she reads from what the screen presents to her. The two other girls involved with the Primer—Elizabeth and Fiona—respond differently to the Primer, and the text attributes these different responses to the human performers linked to the particular copies of the Primer. John Hackworth ractors (performs) for his daughter Fiona. Like Nell, she develops a personal relationship with the Primer. On the other hand, Elizabeth, whose ractors keep changing based on availability, loses interest in the Primer. The human factor, or human nurturing, as shown in the role of the human performer, seems to be an important aspect of the effectiveness of the device, which in a way conveys the author's anxiety about defining the role of humans in this human–machine coupling.

The text thus contains two intertwined narratives, the real-life narrative of the main characters and the intertwined virtual narrative of Princess Nell, which becomes Nell's virtual persona. In the cybernetic circuit of hybrid subjectivity and desires, the embodied reality of Nell and the represented reality of Princess Nell come together. It is at the point at which the two become one in Nell's mind that her quest enters its final phase in resurrecting her surrogate mother Miranda from the Society of Drummers and in becoming the leader of a quarter of a million Han girls who have followed Nell's adventures for years through their copies of the Primer.

The Diamond Age presents various constructions of humans—humans as machines, machines as humans, and human–machine coupling. In a world permeated with simulations, how is one to know who is behind the simulations—the intelligent machines, machinelike humans, or humans with ethical values? Humans as machines are represented by the Society of Drummers, who live in an underwater system of tunnels. Due to injected nanocytes, they live in a trance state in which only the computational circuits of their brains work. The nanoparticles, carrying bits of information, travel among the Drummers through the exchange of bodily fluids during an orgiastic ritual, turning the participants into a powerful computing machine. The humanity of the Drummers is thus reduced to the information-processing function of their collective hive brain, which works as a code-breaking device. They serve an important function; it is through joining the Drummers that John Hackworth develops various components of his Seed project. Also, Miranda joins the Society of Drummers in order to use the code-breaking ability of their hive brain to find out the whereabouts of Nell. The text thus acknowledges the importance of humans working as machines to discover new ways to solve problems, even though Stephenson wants to keep the distinction between humans and machines intact.

Nell's virtual journey is an apprenticeship in learning to distinguish humans from machines in order to find out who is behind the simulations she encounters in different virtual worlds. In her journey through the Kingdom of Turing, Nell realizes that the major difference between humans and machines is that humans have presence as conveyed through their emotional response to situations and people and their capability for caring for one another. That is not to say that all humans continue to retain their presence, as is seen in the case of Nell's biological mother, who is incapable of asserting her presence and is more absent than present in her daughter's life because she spends hours hooked to the mediatron. The text concedes that technical rationality as exhibited by Turing machines who pretend to be human or humans who act as machines is not enough to end human subjugation.

The theme of Nell's education through the Primer is intertwined with that of the revolt by the ethnic Chinese against the foreigners as well as the imported western culture that has made a small percentage prosperous while the majority are suffering deprivation and death. According to Dr. X of the Celestial Kingdom, the importation of western technology into China has brought with it a western worldview that promotes the existence of a few major producers while the rest of the population is turned into consumers. In the Coastal Republic, commercialism is rampant. Even skyscraper walls have become billboards to display mediatronic images of consumer goods and entertainment services in all colors and shapes. An increasing number of poor people are pushed to the fringes both geographically and economically. In the deep interior of China, peasants are dying of starvation and paddy fields are barren because people prefer synthetic rice from the Feed rather than the rice from the paddy grown in the fields. The Feed technology is primarily controlled by the Victorians, who own the Source that supplies a stream of molecules distributed through a Feed network to privately owned Matter Compilers that produce food as well as other daily needs. Other tribes, such as the Nippon, also control some Feed networks. The foreign Feed networks have resulted in changing the dynamics of Chinese society, because increasing numbers of people have stopped planting their fields.

Dr. X recognizes the power of the Seed technology to help the Chinese people shake off the shackles of the foreign Feed, which has destroyed their traditional values and brought chaos and disorder to their society. He dreams of a future for his people in which there will be many producers and very few consumers. The Common Economic Protocol guarding the economic interests of major tribes is against the Seed technology because it can easily be misused to make weapons. In reality, however, their opposition is grounded in the fact that it will undermine the economic power of the major tribes by empowering the vast majority, who would no longer need the Feed. Dr. X is not concerned about the misuse of the Seed technology.

In the Celestial Kingdom based on Confucian ethics, he says, self-discipline is a valued virtue. If an individual is self-disciplined, then "the family is orderly, the village is orderly, the state is orderly. In our hands the Seed would be harmless" (416). Dr. X subversively promotes further development of the Seed technology as he recruits the nanoengineer John Hackworth, without his knowledge, to work with the underwater Society of Drummers for ten years. When Hackworth emerges from this assignment and eventually finds out the nature of his mission, he confronts Dr. X and refuses to cooperate with him to bring the project to completion because the Victorian tribe to which he owes his allegiance is against this technology. The Seed project, however, is not dead; the text points to Nell as the person who will turn this technology into reality.

At the end of her virtual adventure in the Primer, Nell meets the alchemist behind the amazing structures of the Land Beyond, a nano-technological wonder. And the alchemist is King Coyote, the virtual persona of John Hackworth, who hands over all his books, including the Book of the Seed, to Nell as she finishes her virtual quest, which also marks the end of the Primer. The same direction of the action is also hinted at by the image that Nell displays on the outside walls of the skyscraper where she is trapped in an attack by the radical Confucian group called the Fists, who have risen in revolt to end the tyranny of the foreigners controlling the Feed technology. The image she draws replaces the advertising between the hundredth and the two hundredth floors of the skyscraper with "a simple line drawing in primary colors: an escutcheon in blue, and within it, a crest depicting a book drawn in red and white; crossed keys in gold; and a seed in brown" (434). This image becomes a signal to the army of young Han girls who are marching toward the area in search of their Queen Nell, whom they have come to know through their copies of the Primer.

In the future imagined by Stephenson in *The Diamond Age*, the best way to educate young people in a highly commercial and technically advanced society is through interactive devices that make use of archetypal figures from various cultural traditions to create simulations that the users act out to be trained in moral and ethical behavior. The text presents a clash between two cultures, neo-Victorian and neo-Confucian. The Victorian model is depicted as based on rational and scientific principles and is conducive to a society in which the members know their places in the system and respect the rules of the game. The Confucian model is presented as excelling in filial piety and love of nature, as embodied in the image on Dr. X's calling card, which depicts a man with a conical hat slung on his back sitting on a rock with a bamboo pole hauling a fish or a dragon out of the water. The text thus presents a tension not only between two different paths of technological development, the Feed and the Seed technologies, but also two different modes of social behavior, individualism and communalism. John Hackworth embodies the former in his single-minded devotion to nanotechnology research without any concern for the social implications of his research. Dr. X personifies the latter, as is reflected in his readiness to offer refuge on his ships to hundreds of thousands of Han baby girls and his effort to get them educated. Nell fits somewhere in between, because she exhibits individualism and a rational-scientific side in her aspiration to master technology to achieve her individual goals but at the same time cares for all those who have helped her on her journey, including her surrogate mother. She also becomes a leader of the army of Han girls, who are trained in the Confucian worldview through their copies of the Primer. The ending hints that Nell, who believes that life is full of contradictions and ambiguities, is the one who has the potential of bringing change to China through the wise deployment of the democratic Seed technology.

The theme of educating young people through unconventional means reflects the author's view that we need to go beyond technocratic logic to explore past narratives that are more ecologically sustainable and benefit the majority of people, even as it reiterates that technocratic logic is important for future technological development. The historical models that the text promotes are Victorian and Confucian, which translate into the rational-scientific western worldview and the ecologically sustainable eastern worldview in a globalized, borderless world in which both people and goods flow back and forth and one's class and belief system determine one's identity. In such a world, people who live in different parts of the world can feel part of a worldwide community of like-minded people. The elite, translated as multinationals, who own the means of production, can operate from any part of the world, in the west or the east, to keep capital continuously in circulation. Though the text presents negative aspects of commercialization and the need to go beyond technocratic logic to create a better society, there is no indication of a need for any concrete political change to eliminate oppressive mechanisms of society from the ground up. No new technology can be seen as existing outside the political-economic framework of a culture. The notion of succeeding with the deployment of a decentralized technological innovation in an authoritarian and hierarchical society modeled after Confucian ethics is doubtful. Any ethical and moral education of the younger generation must go hand in hand with a society's acceptance of public rights to their bodies and their environment.

It is ironic that Stephenson envisions the freeing of China from the clutches of western technology through the creation of the Seed technology based on nanotechnology when it is obvious that the current Seed technology, based on genetic engineering, is leading to the disempowerment of millions of farmers in the emerging economies of China and India. In the past, farmers relied on cultivating, storing, and exchanging their own seeds, which preserved both biodiversity and cultural diversity. With the entry of the transnational corporations into the seed industry, seed cultivation has become intimately linked to intellectual property rights. It is in the interest of corporations that develop new varieties of seeds to patent them and make them available to farmers at a high price. Farmers cannot save, store, or exchange such seeds because those actions will infringe on the corporations' intellectual property rights. Also, governments introduce legislation that prevents farmers from storing, exchanging, or using native seeds. Thus farmers who were once producers of their own seed supply become consumers of seeds produced by the seed industry. A booming economy in both China and India has resulted in changing food needs, and the peasant- and farmer-based agriculture is quickly turning into agribusiness-run industrial agriculture controlled by multinational corporations that select what is planted, where it is planted, who consumes it, and at what price.

The research in agricultural biotechnology to produce trans-genic plants is controlled by the multinational corporations, which determine all aspects of this industry. Instead of being dependent on millions of independent farmers, the food industry is quickly giv-ing way to select suppliers of the agribusiness industry that follow the agendas of their buyers. The terminator technology for produc-ing seed adds another twist to this tale, because it involves pro-ducing plants that themselves produce sterile seeds, which means that these second-generation seeds cannot be used, making farmers totally dependent on the multinational corporations from crop to crop. This technology, not yet commercially available, has a poten-tial to strengthen the seed monopolies because such seeds come with no expiration dates and hence are more effective in maintain-ing monopolies compared to the patenting of seeds. Vandana Shiva, a noted Indian environmentalist, writes that it is essential to use the farmers' community rights as opposed to the monopolists' intel-lectual property rights, which they use as justification for privately owning and profiting from public commons. This is destroying the lives and livelihoods of millions of farmers. In view of the reality of what is currently happening in both China and India, Stephenson's vision of change seems a bit naïve in that a desirable future for the majority of people in China cannot be realized by simply reverting to the traditional narratives, adopting successful social models from the past, or developing the Seed technology. The politics of human rights, the labor force, and the maintenance of sustainable and eco-logical practices that take into consideration the well-being of the people and the environment are absolutely necessary to bring about positive change in the world for generations to come.

Ecological consciousness is necessary to resist the technocratic logic that is resulting in the destruction of both humans and the envi-ronment, though it seems to be justified by dominant technocratic forces as a price that the world has to pay for prosperity—a pros-perity that is having a negative impact on the majority of cultures as well as the environment throughout the world. The scientists, technologists, governments, and activist groups in the newly emerging economies, like China and India, need to critically evaluate the practices of modern science and technology as they identify what should be introduced from the outside and what should be produced locally. The important issues for such countries should be finding ways and means of deploying modern scientific and technological developments while retaining traditional means of knowledge that are more ecologically sustainable. New scientific and technological applications should not completely replace traditional modes of knowledge but rather should complement them, because more often than not traditional methods might be more ecologically sustainable and benefit larger groups of people, especially in those countries with large rural populations. For this to happen, an alternative science and technology policy is needed that is sensitive to local cultures and local knowledges. This would, of course, require democratization of both the production as well as the application of scientific and technological knowledge so it can benefit the majority of the population with least damage to the environment.

Advances in science and technology are intricately woven with the economic activities of our culture. The current labor processes, as well as science and technology as institutions, are structured in ways that lead to the domination of people and nature in the eternal quest for efficiency and profits. Consequently, the quality of both human life and the environment is subject to degradation. In order to ensure long-term global survival, ecological consciousness is essential, especially as the capitalistic forces of development destroy vast spaces of the globe in their drive to put both the land and resources into an unending cycle of production and consumption. Many are turning back toward traditional values to find what is missing in a highly consumer-oriented culture. This turning back is not a nostalgia for the return of primitive or feudal modes of being. It is rather a search for holistic values that cannot be materialized through a purely rationalist discourse of modern science and technology. Surprisingly, writers from the First World see the forward progress of society on the technocratic path as an inevitable movement of events, and even when nontechnocratic alternatives are envisioned, they seem to be merely a reshuffling of the old historical scenarios rather than a genuine new direction in society. Critically aware writers who write from the peripheries, however, are more genuine in their expression of the need for change as well as what that change should involve to set humankind on a more democratic course so the oppressed segments of society can be uplifted to better lives and a full flowering of their potential.

NOTES

Introduction

1 In *Postcolonial Aura* Arif Dirlik implicates postcolonial discourse in promoting multiculturalism and diversity and thus serving the interests of global capitalism, in which these concepts are used as strategic tools to create new markets.

2 Paul Connerton writes that the social memory of a community is conveyed and sustained through ritual performances as manifested in rites, rituals, and commemorative ceremonies. The myths and rituals encode the cultural values of a community, and in their ritual enactment these values become part of the lived experience of the community.

3 Gayatri Spivak (1988b) argues that the notion of an authentic culture arises from the colonial constructions in which the other of the western dominant culture is seen to constitute a monolithic entity that exists in purity. Such colonial constructions are repeated in a postcolonial state in which the dominant or mainstream culture is at odds with indigenous cultures.

4 N. Katherine Hayles provides an excellent description of interrelations between scientific models and literary metaphors in *The Cosmic Web: Scientific Field Models and Literary Strategies in the Twentieth Century*.

5 Brian McHale (1992) describes the dominance of modernist fiction as epistemological. By that he means that modernist writers were primarily concerned with epistemological questions regarding the fictional worlds that they created.

6 In a broad analysis of the condition of postmodernity, David Harvey explores the links between changing modes of production and consumption and experimentation in art, literature, and architecture.

7 In an interesting study of fictional space, Carl Malmgren follows the transformation of narrative space from realist to modernist and finally to postmodernist fiction. The realist novel of the eighteenth century and the first half

of the nineteenth displays the writer's confidence in the ability to describe the fictional reality objectively and accurately. The writer or narrator is the master of the fictional domain. Occupying a privileged position with respect to the story told, the writer or narrator appears to have a total knowledge of the world as well as human nature. This confidence in the role of the writer to accurately represent what is going on in the minds of the characters as well as in the world around them is lost in modernist works, which show an increasing disbelief in any totalizing perspectives. The narrator now resists the interpretive and mediating function of earlier fiction and instead retreats into the background and lets the fiction unfold without any commentary. The result is a psychological or stream-of-consciousness novel. As modernist culture leads to postmodern culture, there is yet another transformation of the fictional space. The writer now feels compelled to convey not only the uncertainty of living in a complex, chaotic world but also the disbelief in any metanarratives or totalizing systems of knowledge. Accordingly, in postmodern narrative the fictional space is broken up in order to further open up the narrative space through fragmentation, discontinuity, and self-reflexivity.

8 Morrison's use of black music as a model in her writing implies that the act of listening is important in her writing both from a thematic perspective as well as from the perspective of the reader's participation in co-creating the text with her. Joanna Wolfe notes that in the act of listening to others' stories, the characters in *Beloved* (Denver, Paul D, and Bodwin) reinvent their past texts, which had kept them trapped in logocentric closure with no lines of escape. Through this rewriting and reinventing of the closed text, they become subjects of their own stories.

9 Lynda Koolish comments on the necessity of the trope of memory "in the context of an entire people deliberately denied literacy—allowed no letters, no diaries, no poems, no record of the past—whose events must be understood in order to make sense of the present" (427).

10 The oral quality of black tradition is captured by using narrative strategies to "make the story appear oral, meandering, effortless, spoken—to have the reader feel the narrator without identifying that narrator, or hearing him or her knock about, and to have the reader work with the author in the construction of the book" (Morrison 1984, 341).

11 William Handley writes that in *Beloved* the writer "makes the reader aware of how the acts of writing and reading inevitably involve an ethical imperative, made all the more imperative when what is narrated is unspeakable" (688).

12 While commenting on the collective nature of black literature Morrison says that, if true to its spirit, it is invariably political. In her own writing, she said, she is "not interested in indulging myself in some private, closed exercise

of my imagination that fulfills only the obligation of my personal dreams" (Morrison 1984, 344).

13 J. Yellowlees Douglas (1994a) describes her experience as a reader of *afternoon* and the impact of the lack of closure in the text on the reading experience.

14 Jay Labinger (1995) reads Powers's *Gold Bug* in terms of three coding motifs that appear as computer programming, genetic code, and musical coding.

15 Brian McHale (1992) describes the shift from modernist to postmodernist writing in terms of a shift from the epistemological to the ontological dominant. Postmodern writers are concerned with ontological questions as to how the multiple worlds come into existence, how they exist, and how they collide with and interpenetrate one another.

16 Michel Foucault uses the term "heterotopia" in thinking of the contemporary disordered space of knowledge and experience that is created when fragments from different orders of reality come together in one dimension. Each fragment is pulled out of its context and transplanted into an alien space alongside other fragments that have also been deprived of their original context. Thus there is no common underlying foundation that holds the fragments together as they enter into a dynamic relationship with one another. The disordered or fragmented space of experience is conveyed in postmodern literary works through the use of hypertext or collage.

3. Multiplicity

1 New technologies of representation, Bolter and Grusin write, are shaped by twin logics of mediation: the logic of immediacy and the logic of hypermediacy. The immediacy in film and television is created through giving viewers an "out of the window" view of the scene, as is done in reality shows. Immediacy in digital media is created through interactivity as users assert their active presence by moving around, clicking on links to see images and video clips, or hear sounds. The hypermediated space thus resists a singular gaze, as viewers "do not look *through* the medium in linear perspective; rather, [they] look *at* the medium or at a multiplicity of the media that may appear in windows on a computer screen or in the fragmented elements of collage or a photomontage" (81).

2 In a private communication, Strickland noted that the publishing material was meant to be printed on a shrink wrapper, but Penguin does not shrink-wrap its books.

4. Assemblage

1 Karsten Harries interprets cubist collage as an experimentation with the frame. Broken frames, no frames, or frames absorbed partially or completely

in the field of representation represent a prelude to a turn away from the mimetic function of art because it brings the viewer's attention to the work's autonomy. Harries links the use of a broken frame to the broader state of postmodern culture, which has lost faith in metanarratives of all kinds. A postmodern interpretation, however, shifts the focus from any search for ground in collage to exploring how the multiple worlds depicted in a collage relate to one another and what that says about subjectivity, experience, and the world we live in.

5. Technocracy

1 Not only is the human subject conceived of as a set of information patterns, but the human body itself becomes literally a surface to write upon, as is seen in the advances in cosmetic surgery, in which the cosmetic surgeons inscribe the culturally valued signs on the desiring bodies. Drawing upon Haraway's work, Anne Balsamo explores the inscription of gender in technically crafted denaturalized human bodies as she focuses on material practices to recraft and reshape the human body as manifested in the varying cultures of cosmetic surgery and female body building, among others. She writes: "When seemingly stable boundaries are displaced by technological innovations (human/artificial, life/death, nature/culture), other boundaries are more vigilantly guarded. Indeed, the gendered boundary between male and female is one border that remains heavily guarded despite new technological ways to rewrite the physical body in the flesh. So it appears that while the body has been recoded within the discourses of biotechnology and medicine as belonging to an order of culture rather than of nature, gender remains a naturalized marker of human identity" (9).

BIBLIOGRAPHY

Aarseth, Espen J. 1997. *Cybertext: Perspectives on Ergodic Literature.* Baltimore, Md.: Johns Hopkins University Press.

Abish, Walter. 1974. *Alphabetical Africa.* New York: New Directions Pub Corp.

Abrams, M. H. 1953. *The Mirror and the Lamp: Romantic Theory and the Critical Tradition.* New York: Oxford University Press.

Ahmad, A. 1992. *In Theory: Classes, Nations, Literatures.* Oxford: Oxford University Press.

Anzaldúa, Gloria. 1987. *Borderlands / La Frontera: The New Mestiza.* San Francisco: Spinsters / Aunt Lute.

Appadurai, Arjun. 1996. *Modernity at Large: Cultural Dimensions of Globalization.* Minneapolis: University of Minnesota Press.

Ashcroft, Bill. 2001. *On Post-Colonial Futures: Transformations of Colonial Cultures.* London and New York: Continuum.

Ashcroft, Bill, G. Griffiths, and H. Tiffin. 1989. *The Empire Writes Back: Theory and Practice in Postcolonial Literatures.* London: Routledge.

Auster, Paul. 1987. *City of Glass.* New York: Penguin.

Balsamo, Anne. 1996. *Technologies of the Gendered Body: Reading Cyborg Women.* Durham, N.C.: Duke University Press.

Barth, John. 1988. *Lost in the Funhouse.* New York: Anchor Press.

Barthelme, Donald. 1987. *Forty Stories.* New York: Putnam.

Baudrillard, Jean. 1988. *The Ecstasy of Communication.* Trans. Bernard Schutze and Caroline Schutze. New York: Semiotexte.

Benjamin, Walter. 1968. *Illuminations: Essays and Reflections.* Ed. Hannah Arendt. New York: Schocken Books.

Bhabha, Homi K. 1994. *The Location of Culture.* London: Routledge.

———. 1995. "Cultural Diversity and Cultural Differences." In *PostColonial Studies Reader,* ed. Bill Ashcroft, Gareth Griffiths, and Helen Tiffin, 206–9. London and New York: Routledge.

Boden, Margaret A., ed. 1990. *The Philosophy of Artificial Intelligence.* Oxford: Oxford University Press.

Bois, Yves-Alain. 1993. *Painting as Model.* Cambridge, Mass.: MIT Press.

Bolter, Jay David. 1991. *Writing Space: The Computer Hypertext and the History of Writing.* Hillsdale, N.J., and London: Lawrence Erlbaum Associates.

———. 2001. *Writing Space: Computers, Hypertext, and the Remediation of Print.* 2nd ed. Mahwah, N.J., and London: Lawrence Erlbaum Associates.

Bolter, Jay David, and Richard Grusin. 1999. *Remediation: Understanding New Media.* Cambridge, Mass.: MIT Press.

Borges, Jorge Louis. 1962a. "The Garden of Forking Paths." In *Labyrinths,* ed. Donald A. Yates and James E. Irby, 19–29. New York: New Directions.

———. 1962b. *Labyrinths.* Ed. Donald A. Yates and James E. Irby. New York: New Directions.

Brockelman, Thomas P. 2000. *Frame and the Mirror: On Collage and the Postmodern.* Evanston, Ill.: Northwestern University Press.

Bukatman, Scott. 1993. *Terminal Identity: The Virtual Subject in Postmodern Science Fiction.* Durham, N.C.: Duke University Press.

Butler, Judith. 1993. *Bodies That Matter.* London: Routledge.

Calvino, Italo. 1968. *Cosmicomics.* Trans. William Weaver. New York: Harcourt Brace Jovanovich.

———. 1969. *t zero.* Trans. William Weaver. New York: Harcourt Brace Jovanovich.

———. 1974. *Invisible Cities.* Trans. William Weaver. New York: Harcourt Brace.

Carroll, Lewis. 1993. *Alice's Adventures in Wonderland.* London: Dent.

———. 2003. *Alice's Adventures in Wonderland and Through the Looking-Glass.* London: Penguin Classics.

Chartier, Roger. 1994. *The Order of Books.* Trans. Lydia G. Cochrane. Stanford, Calif.: Stanford University Press.

Chow, Rey. 1993. *Writing Diaspora: Tactics of Intervention in Contemporary Cultural Studies.* Bloomington: Indiana University Press.

———. 1995. *Primitive Passions.* New York: Columbia University Press.

Ciccoricco, David. 2007. *Reading Network Fiction.* Tuscaloosa: University of Alabama Press.

Cixous, Hélène. 1981. *Readings: The Poetics of Blanchot, Joyce, Kafka, Kleist, Lispector, and Tsvetayeva.* Minneapolis: University of Minnesota Press.

Connerton, Paul. 1989. *How Societies Remember.* New York: Cambridge University Press.

Coover, Robert. 1969. *Pricksongs & Descants: Fictions.* New York: Plume.

Cortázar, Julio. 1966. *Hopscotch.* Trans. Gregory Rabassa. New York: Random House.

Coverley, M. D. 2001. *Califia.* Cambridge, Mass.: Eastgate Systems.

———. 2006. *Egypt: The Book of Going Forth by Day*. Newport Beach, Calif.: Horizon Insight.

Cutter, Martha J. 2000. "The Story Must Go On and On: The Fantastic, Narration, and Intertextuality in Toni Morrison's *Beloved* and *Jazz*." *African American Review* 34, no. 1: 61–75.

Danielewski, Mark Z. 2000. *House of Leaves*. New York: Pantheon Books.

Danto, Arthur. 1997. *After the End of Art: Contemporary Art and the Pale of History*. Princeton, N.J.: Princeton University Press.

Dawkins, Richard. 1976. *The Selfish Gene*. New York: Oxford University Press.

de Certeau, Michel. 1987. *The Practice of Everyday Life*. Berkeley and Los Angeles: University of California Press.

Deleuze, Gilles. 1986. *Cinema 1: The Movement Image*. Trans. Hugh Tomlinson and Barbara Habberjam. Minneapolis: University of Minnesota Press.

———. 1989. *Cinema 2: The Time-Image*. Trans. Hugh Tomlinson and Robert Gaaleta. Minneapolis: University of Minnesota Press.

———. 1990. *The Logic of Sense*. Trans. Mark Lester with Charles Stivale. Ed. Constantin V. Boundas. New York: Columbia University Press.

———. 1993. *The Fold: Leibniz and the Baroque*. Trans. Tom Conley. Minneapolis: University of Minnesota Press.

———. 1994. *Difference and Repetition*. Trans. Paul Patton. New York: Columbia University Press.

———. 1997. *Essays Critical and Clinical*. Trans. Daniel B. Smith and Michael A. Greco. Minneapolis: University of Minnesota Press.

———. 2000. *Proust and Signs: The Complete Text*. Trans. Richard Howard. Minneapolis: University of Minnesota Press.

Deleuze, Gilles, and Félix Guattari. 1986. *Franz Kafka: Toward a Minor Literature*. Trans. Dana Polan. Minneapolis: University of Minnesota Press.

———. 1987. *A Thousand Plateaus: Capitalism and Schizophrenia*. Trans. Brian Massumi. Minneapolis: University of Minnesota Press.

———. 1996. *Anti-Oedipus: Capitalism and Schizophrenia*. Trans. Robert Hurley, Mark Seem, and Helen R. Lane. Minneapolis: University of Minnesota Press.

Dennett, Daniel C. 1992. *Consciousness Explained*. Boston: Back Bay Books.

de Visscher, Eric. 1993. "'There's No Such a Thing as Silence . . .': John Cage's Poetics of Silence." In *Writings about John Cage*, ed. Richard Kostelanetz, 117–33. Ann Arbor: University of Michigan Press.

Dirlik, Arif. 1994. *After the Revolution: Waking to Global Capitalism*. Hanover, N.H.: Wesleyan University Press.

———. 1997. *The Postcolonial Aura: Third World Criticism in the Age of Global Capitalism*. Boulder, Colo.: Westview.

Dobbs, Cynthia. 1998. "Toni Morrison's *Beloved:* Bodies Returned, Modernism Revisited." *African American Review* 32, no. 4: 563–78.

Douglas, J. Yellowlees. 1994a. "'How Do I Stop This Thing?' Closure and Indeterminacy in Interactive Narratives." In *Hyper/Text/Theory,* ed. George P. Landow, 159–88. Baltimore, Md.: Johns Hopkins University Press.

———. 1994b. *I Have Said Nothing.* Cambridge, Mass.: Eastgate Systems.

———. 2001. *The End of Books—or Books without End? Reading Interactive Narratives.* Ann Arbor: University of Michigan Press.

Doyle, Richard. 1997. *On Beyond Living: Rhetorical Transformations in the Life Sciences.* Stanford, Calif.: Stanford University Press.

Dreyfus, Hubert L. 1979. *What Computers Can't Do: The Limits of Artificial Intelligence.* Rev. ed. New York: Harper and Row.

Edelman, Gerald M. 1992. *Bright Air, Brilliant Fire.* New York: Basic Books.

Egyptian Book of the Dead. 2008. Trans. Wallace Budge. Ed. John Romer. New York: Penguin.

Eisenstein, Elizabeth. 1983. *The Printing Revolution in Early Modern Europe.* New York: Cambridge University Press.

Eisenstein, Sergei. 1985. "From Film Form: 'The Cinematographic Principle and Ideogram: A Dialectical Approach to Film Form.'" In *Film Theory and Criticism,* ed. Gerald Mast and Marshall Cohen, 90–123. Oxford: Oxford University Press.

Eskelinen, Markku. 2001. "Cybertext Theory and Literary Studies." *Electronic Book Review* 12. http://www.electronicbookreview.com/ebr12/eskel.htm (accessed April 2002).

Fanon, Frantz. 1967. *Black Skins, White Masks.* Trans. Charles Lam Markmann. New York: Grove Press.

———. 1990. *The Wretched of the Earth.* Trans. Constance Farrington. New York: Penguin.

Federman, Raymond. 1993. *Critifiction: Postmodern Essays.* New York: State University of New York Press.

Feenberg, Andrew. 2002. *Transforming Technology: A Critical Theory Revisited.* New York: Oxford University Press.

Fisher, Caitlin. 2001. *These Waves of Girls.* http://www.yorku.ca/caitlin/waves/ (accessed November 2005).

Foster, Thomas. 1999. "'The Souls of Cyber-Folk': Performativity, Virtual Embodiment and Racial Histories." In *Cyberspace Textuality: Computer Technology and Literary Theory,* ed. Marie-Laure Ryan, 137–63. Bloomington: Indiana University Press.

Foucault, Michel. 1970. *The Order of Things: An Archeology of the Human Sciences.* New York: Pantheon.

Francese, Joseph. 1997. *Narrating Postmodern Time and Space.* New York: State University of New York Press.

Franklin, Stan. 1997. *Artificial Minds.* Cambridge, Mass.: MIT Press.

Gaggi, Silvio. 1997. *From Text to Hypertext.* Philadelphia: University of Pennsylvania Press.

Gandhi, Leela. 1998. *Postcolonial Theory: A Critical Introduction.* New York: Columbia University Press.

Gass, William H. 1971. *Willie Master's Lonesome Wife.* New York: Alfred A. Knopf.

Gates, Henry Louis, Jr., ed. 1986. *"Race," Writing, and Difference.* Chicago: University of Chicago Press.

Gibson, William. 1984. *Neuromancer.* New York: Ace Books.

Gilbert, Sandra M., and Susan Gubar. 1984. *The Madwoman in the Attic: The Woman Writer and the Nineteenth Century Literary Imagination.* New Haven, Conn.: Yale University Press.

Gins, Madeline. 1969. *Word Rain.* New York: Grossman Publishers.

Guyer, Carolyn. 1992. *Quibbling.* Watertown, Mass.: Eastgate Systems.

———. 1993. "Introduction." In *its name was Penelope* by Judy Malloy. Cambridge, Mass.: Eastgate Systems.

Guyer, Carolyn, and Martha Petry. 1991a. *Izme Pass,* a collaborative hyperfiction. In *Writing on the Edge* 2, no. 2. Bound-in computer disk, University of California at Davis.

———. 1991b. "Notes for Izme Pass Expose." In *Writing on the Edge* 2, no. 2: 82–89.

Halberstam, Judith, and Ira Livingston. 1995. *Posthuman Bodies.* Bloomington: Indiana University Press.

Handley, William R. 1995. "'The House a Ghost Built': Nommo, Allegory, and the Ethics of Reading in Toni Morrison's *Beloved." Contemporary Literature* 36: 676–701.

Haraway, Donna. 1991. *Simians, Cyborgs, and Women: The Reinvention of Nature.* New York: Routledge.

Harding, Sandra, ed. 1993. *The "Racial Economy of Science": Toward a Democratic Future.* Bloomington: Indiana University Press.

Harpold, Terence. 1991. "The Contingencies of the Hypertext Link." *Writing on the Edge* 2, no. 2: 126–38.

Harries, Karsten. 1989. *The Broken Frame: Three Lectures.* Washington, D.C.: Catholic University Press of America.

Harris, Paul A. 1993. "Epistemocritique: A Synthetic Matrix." *SubStance: A Review of Theory and Criticism* 22, no. 71/72: 185–203.

Harvey, David. 1989. *The Condition of Postmodernity: An Enquiry into the Origins of Cultural Change.* New York: Blackwell Publishers.

Hayles, N. Katherine. 1984. *The Cosmic Web: Scientific Field Models and Literary Strategies in the Twentieth Century*. Ithaca, N.Y.: Cornell University Press.

———. 1990. *Chaos Bound: Orderly Disorder in Contemporary Literature and Science*. Ithaca, N.Y.: Cornell University Press.

———. 1993. "Virtual Bodies and Flickering Signifiers." *October* 66:69–91.

———. 1997. "The Posthuman Body: Inscription and Incorporation in *Galatea 2.2* and *Snow Crash*." *Configurations* 5:241–66.

———. 1999. *How We Became Posthuman: Virtual Bodies in Cybernetics, Literature, and Informatics*. Chicago: University of Chicago Press.

———. 2001. "Cyber/literature and Multicourses: Rescuing Electronic Literature from Infanticide." Riposte to Montfort. *Electronic Book Review* 11. http://www.altx.com/ebr/riposte/rip11/rip11hay.htm (accessed November 2005).

———. 2002. *Writing Machines*. Cambridge, Mass.: MIT Press.

———. 2003. "Deeper into the Machine: The Future of Electronic Literature." *Culture Machine* 5. http://culturemachine.tees.ac.uk/Cmach/Backissues/j005/Articles/Hayles/NHayles.htm (accessed November 2005).

———. 2005. *My Mother Was a Computer: Digital Subjects and Literary Texts*. Chicago: University of Chicago Press.

———. 2008. *Electronic Literature: New Horizons for the Literary*. Notre Dame, Ind.: University of Notre Dame Press.

———, ed. 2004. *Nanoculture: Implications of the New Technoscience*. Portland, Ore.: Intellect Books.

Heim, Michael. 1998. *Virtual Realism*. Oxford: Oxford University Press.

Horvitz, Deborah. 1990. "Nameless Ghosts: Possession and Dispossession in *Beloved*." *Studies in American Literature* 77:157–67.

House, Elizabeth B. 1990. "Toni Morrison's Ghost: The Beloved Who Is Not Beloved." *Studies in American Fiction* 18:17–26.

Hutcheon, Linda. 1988. *A Poetics of Postmodernism: History, Theory, Fiction*. New York: Routledge.

———. 1989. *The Politics of Postmodernism*. New York: Routledge.

Hutchins, Edwin. 1995. *Cognition in the Wild*. Cambridge, Mass.: MIT Press.

Irigaray, Luce. 1985. *This Sex Which Is Not One*. Trans. Catherine Porter with Carolyn Burke. Ithaca, N.Y.: Cornell University.

———. 1993. *An Ethics of Sexual Difference*. Trans. Carolyn Burke and Gillian C. Gill. Ithaca, N.Y.: Cornell University Press.

———. 1994. *Speculum of the Other Woman*. Trans. Gillian C. Gill. Ithaca, N.Y.: Cornell University Press.

Jackson, Shelley. 1995. *Patchwork Girl*. Cambridge, Mass.: Eastgate Systems.

Jameson, Fredric. 1991. *Postmodernism, or The Cultural Logic of Late Capitalism*. Durham, N.C.: Duke University Press.

Johnson, Steven. 1997. *Interface Culture: How New Technology Transforms the Way We Create and Communicate*. New York: Basic Books.

Joyce, James. 1984. *Ulysses*. New York: Garland.

Joyce, Michael. 1987. *afternoon, a story*. Cambridge, Mass.: Eastgate Systems.

———. 1991. *WOE—A Memory of What Will Be*. In *Writing on the Edge* 2, no. 2.

———. 1995. *Of Two Minds: Hypertext Pedagogy and Poetics*. Ann Arbor: University of Michigan Press.

———. 1996. *Twilight, a symphony*. Watertown, Mass.: Eastgate Systems.

———. 2000. *Othermindedness: The Emergence of Network Culture*. Ann Arbor: University of Michigan Press.

Katz, Steve. 1968. *The Exagggerations [sic] of Peter Prince: The Novel*. New York: Holt, Rinehart and Winston.

Kittler, Friedrich A. 1990. *Discourse Networks 1800/1900*. Trans. Michael Metter with Chris Cullens. Stanford, Calif.: Stanford University Press.

Koolish, Lynda. 1995. "Fictive Strategies and Cinematic Representations in Toni Morrison's *Beloved:* Postcolonial Theory / Postcolonial Text." *African American Review* 29, no. 3: 421–38.

Krauss, Rosalind E. 1986. *The Originality of the Avant Garde and Other Modernist Myths*. Cambridge, Mass.: MIT Press.

Kuhn, Annette. 1985. *The Power of the Image*. London: Routledge, Kegan Paul.

Labinger, Jay A. 1995. "Encoding an Infinite Message: Richard Powers' *Gold Bug Variations*." *Configurations* 3, no. 1: 79–93.

Landow, George P. 1993. *Hypertext: The Convergence of Contemporary Critical Theory and Technology*. Baltimore, Md.: Johns Hopkins University Press.

———. 1999. "Hypertext as Collage Writing." In *The Digital Dialectic: New Essays on New Media*, ed. Peter Lunenfeld, 150–70. Cambridge, Mass.: MIT Press.

Lanham, Richard A. 1993. *The Electronic Word: Democracy, Technology, and the Arts*. Chicago: University of Chicago Press.

Latour, Bruno. 1986. *Laboratory Life*. Princeton, N.J.: Princeton University Press.

———. 1987. *Science in Action: How to Follow Scientists and Engineers through Society*. Cambridge, Mass.: Harvard University Press.

———. 1997. *We Have Never Been Modern*. Trans. Catherine Porter. Cambridge, Mass.: Harvard University Press.

———. 1999. *Pandora's Hope*. Cambridge, Mass.: Harvard University Press.

Lefebvre, Henri. 1994. *The Production of Space*. Trans. Donald Nicholson-Smith. Oxford: Blackwell Publishers.

Luesebrink, Marjorie. N.d. "Artists and Innovators: An Interview with Marjorie Luesebrink" by Thomas Swiss. http://popmatters.com (accessed November 2005).

Lunenfeld, Peter, ed. 1999. *The Digital Dialectic: New Essays on New Media*. Cambridge, Mass.: MIT Press.

Lyotard, Jean-François. 1984. *The Postmodern Condition*. Minneapolis: University of Minnesota Press.

Malloy, Judy. 1993. *its name was Penelope*. Cambridge, Mass.: Eastgate Systems.

———, ed. 2004. *Women, Art, and the Media*. Cambridge, Mass.: MIT Press.

Malmgren, Carl D. 1985. *Fictional Space in the Modernist and Postmodernist American Novel*. Lewisburg, Penn.: Bucknell University Press.

Manovich, Lev. 2001. *The Language of New Media*. Cambridge, Mass.: MIT Press.

McHale, Brian. 1987. *Postmodernist Fiction*. New York: Methuen.

———. 1992. *Constructing Postmodernism*. London: Routledge.

Mckay, Nellie Y., ed. 1988. *Critical Essays on Toni Morrison*. Boston: G. K. Hall.

McLuhan, Marshall. 1992. *The Gutenberg Galaxy*. Toronto: University of Toronto Press.

Memmott, Talan. 2000. *Lexia to Perplexia*. http://www.uiowa.edu/~iareview/tirweb/hypermedia/talan_memmott/ (accessed November 2005).

Milburn, Colin. 2002. "Nanotechnology in the Age of Posthuman Engineering: Science Fiction as Science." *Configurations* 10:261–95.

Minsky, Marvin. 1985. *The Society of Mind*. New York: Simon and Shuster.

Mitchell, William J. 1994. *The Reconfigured Eye: Visual Truth in the Post-photographic Era*. Cambridge, Mass.: MIT Press.

Montfort, Nick. 2001. "Cybertext Killed the Hypertext Star." *Electronic Book Review* 11. http://www.electronicbookreview.com/ebr11 (accessed April 2002).

Moore-Gilbert, Bart. 1997. *Postcolonial Theory: Contexts, Practices, Politics*. London: Verso.

Moravec, Hans. 1988. *Mind Children: The Future of Robot and Human Intelligence*. Cambridge, Mass.: Harvard University Press.

Morrison, Toni. 1984. "Rootedness: The Ancestor as Foundation." In *Black Women Writers (1950–1980): A Critical Evaluation*, ed. Mari Evans, 339–45. New York: Anchor Books.

———. 1988. *Beloved*. New York: New American Library.

———. 1989. "Unspeakable Things Unspoken: The Afro-American Presence in American Literature." *Michigan Quarterly Review* 28:1–34.

———. 1990. *Playing in the Dark: Whiteness and the Literary Imagination*. Cambridge, Mass.: Harvard University Press.

———. 1994a. "Interview with Nellie McKay." In *Conversations with Toni Morrison*, ed. Danielle Taylor-Guthrie, 138–45. Jackson: University Press of Mississippi.

———. 1994b. *The Nobel Lecture In Literature, 1993*. New York: Alfred A. Knopf.

———. 1998. "The Site of Memory." Excerpted in *Toni Morrison: Beloved*, ed. Carl Plasa, 43–47. New York: Columbia University Press.

Moulthrop, Stuart. 1991. *Victory Garden*. Watertown, Mass.: Eastgate Systems.

———. 1994. "Rhizome and Resistance: Hypertext and the Dream of a New Culture." In *Hyper/Text/Theory*, ed. George P. Landow, 299–319. Baltimore, Md.: Johns Hopkins University Press.

Murray, Janet H. 1997. *Hamlet on the Holodeck: The Future of Narrative in Cyberspace.* Cambridge, Mass.: MIT Press.

Nakamura, Lisa. 2002. *Cybertypes: Race, Ethnicity, and Identity on the Internet.* New York: Routledge.

Negroponte, Nicholas. 1995. *Being Digital.* New York: Alfred A. Knopf.

Nelson, Theodor Holm. 1987. *Computer Lib/Dream Machines.* U.S.A: Microsoft Press.

———. 1994. *Literary Machines.* Sausalito, Calif.: Mindful Press.

O'Gorman, Marcel. 2006. *E-Crit: Digital Media, Critical Theory, and the Humanities.* Toronto: University of Toronto Press.

Olkowski, Dorothea. 1999. *Gilles Deleuze and the Ruin of Representation.* Berkeley: University of California Press.

Ong, Walter. 1982. *Orality and Literacy: The Technologizing of the Word.* London: Routledge.

Orwell, George. 1992. *Nineteen Eighty-four.* New York: Knopf.

Patton, Paul, ed. 1996. *Deleuze: A Critical Reader.* Cambridge, Mass.: Blackwell Publishers.

Penrose, Roger. 1989. *The Emperor's New Mind.* Oxford: Oxford University Press.

Perez-Torres, Rafael. 1997. "Knitting and Knotting the Narrative Thread—*Beloved* as Postmodern Novel." In *Toni Morrison: Critical and Theoretical Approaches,* ed. Nancy J. Peterson, 91–109. Baltimore, Md.: Johns Hopkins University Press.

Plasa, Carl, ed. 1998. *Toni Morrison, Beloved.* New York: Columbia University Press.

Poe, Edgar Allan. 1991. *The Gold-Bug and Other Tales.* New York: Dover Publications.

Porusch, David. 1994. "Hacking the Brainstem: Postmodern Metaphysics and Stephenson's *Snow Crash.*" *Configurations* 3:537–71.

Postman, Neil. 1992. *Technopoly: The Surrender of Culture to Technology.* New York: Alfred A. Knopf.

Powers, Richard. 1992. *The Gold Bug Variations.* New York: HarperCollins.

———. 1995. *Galatea 2.2.* New York: HarperCollins.

Pratt, Mary Louis. 1986. "Scratches on the Face of the Country, or What Mr. Barrow Saw in the Land of the Bushman." In *"Race," Writing, and Difference,* ed. Henry Louis Gates Jr., 138–62. Chicago: University of Chicago Press.

———. 1992. *Imperial Eyes: Travel Writing and Transculturation.* London: Routledge.

Prigogine, I., and I. Stengers. 1984. *Order out of Chaos.* New York: Bantam Books.

Provenzo, Eugene F. 1986. *Beyond the Gutenberg Galaxy: Microcomputers and the Emergence of Post-typographic Culture.* New York: Teachers College Press.

Queneau, Raymond. 1961. *Cent Mille Miliards de Poemes.* Paris: Gallimard.

Rheingold, Howard. 1991. *Virtual Reality.* New York: Simon and Shuster.

———. 1993. *Virtual Community: Homesteading on the Electronic Frontier.* New York: HarperCollins.

Rifkin, Jeremy. 1998. *Biotech Century: Harnessing the Gene and Remaking the World.* New York: Tarcher and Putnam.

Rodari, Florian. 1988. *Collage: Pasted, Cut, and Torn Papers.* New York: Rizzoli International.

Rorty, Richard. 1980. *Philosophy and the Mirror of Nature.* Princeton, N.J.: Princeton University Press.

Rosello, Mireille. 1994. "The Screener's Maps: Michel de Certeau's 'Wandersmanner' and Paul Auster's Hypertexual Detective." In *Hyper/Text/Theory,* ed. George P. Landow, 121–57. Baltimore, Md.: Johns Hopkins University Press.

Rushdie, Salman. 1995. *Midnight's Children.* London: Vintage.

Ryan, Marie-Laure, ed. 1999. *Cyberspace Textuality: Computer Technology and Literary Theory.* Bloomington: Indiana University Press.

Said, Edward. 1979. *Orientalism.* New York: Vintage.

———. 1993. *Culture and Imperialism.* London: Chatto & Windus.

Shelley, Mary Wollstonecraft. 1984. *Frankenstein, or The Modern Prometheus.* Los Angeles: University of California Press.

———. 1992. "Introduction." In *Frankenstein* by Mary Shelley. Ed. Johanna M. Smith. Boston: St. Martin's Press.

Shiva, Vandana. 2000. *Stolen Harvest: The Hijacking of the Global Food Supply.* Cambridge, Mass.: South End Press.

Shiva, Vandana, and Kunwar Jalees. 2006. *Seeds of Suicide: The Ecological and Human Costs of Seed Monopolies and Globalisation of Agriculture.* New Delhi: Navdanya.

Silko, Leslie Marmon. 1977. *Ceremony.* New York: Penguin.

———. 1981. *Storyteller.* New York: Arcade.

———. 1991. *Almanac of the Dead.* New York: Penguin.

———. 1996. *Yellow Woman and a Beauty of the Spirit.* New York: Simon and Shuster.

Soja, Edward W. 1989. *Postmodern Geographies: The Reassertion of Space in Critical Social Theory.* London: Verso.

Solomon, Barbara H., ed. 1988. *Critical Essays on Toni Morrison's Beloved.* New York: G. K. Hall.

Spivak, Gayatri Chakravorty. 1988a. "Can the Subaltern Speak?" In *Marxism and the Interpretation of Culture,* ed. Cary Nelson and Lawrence Grossberg, 271–313. Chicago: University of Illinois Press.

———. 1988b. *In Other Worlds: Essays in Cultural Politics.* New York: Routledge.

———. 1990. *The Postcolonial Critic: Interviews, Strategies, Dialogues.* Ed. Sarah Harasym. New York: Routledge.

————, trans. and intro. 1995. *Imaginary Maps* by Mahashweta Devi. New York: Routledge.

Stephenson, Neal. 1992. *Snow Crash.* New York: Bantam Books.

————. 1995. *The Diamond Age.* New York: Bantam.

Stone, Allucquere Rosanne. 1995. *The War of Desire and Technology at the Close of the Mechanical Age.* Cambridge, Mass.: MIT Press.

Strickland, Stephanie. 1998. "Seven-League Boots: Poetry, Science, and Hypertext." *Electronic Book Review.* http://altx.com/ebr/ebr7/7strick/ (accessed October 2008).

————. 1999a. "The Ballad of Sand and Harry Soot." Print edition in *Boston Review.* Available at http://bostonreview.mit.eduIBR24. 5/strickland.html (accessed August 2004).

————. 1999b. "Seven Reasons Why Sandsoot Is the Way It Is." *Proceedings of the Cybermountain Colloquium.* Technical Report AUE-CS-99-05, Computer Science Department, Aalborg University, Esbjerg, Denmark. http://www.wordcircuits.com/htww/strickland.htm.

————. 2002a. "Into the Space of Previously Undrawable Diagrams: An Interview with Stephanie Strickland" by Jaishree Odin. *Iowa Web Review.* http://www.uiowa.edu (accessed August 2004).

————. 2002b. *V: WaveSon.nets/Losing L'una.* New York: Penguin.

Strickland, Stephanie, and Janet Holmes. 1999. "The Ballad of Sand and Harry Soot." *Word Circuits.* http://www.wordcircuits.com (accessed August 2004).

Strickland, Stephanie, and Cynthia Lawson. 2002. *V: Vniverse.* http://vniverse.com (accessed November 2005).

————. 2003. "Making the *Vniverse* by Strickland and Lawson." *New River: Journal of Digital Writing and Art* (May). http://www.cddc.vt.edu/journals/newriver/strickland/essay/index.html (accessed August 2004).

Tabbi, Joseph. 1995. *The Postmodern Sublime: Technology and American Writing from Mailer to Cyberpunk.* Ithaca, N.Y.: Cornell University Press.

————. 2001. "The Cybernetic Turn: Literary into Cultural Criticism." *Electronic Book Review* 12. http://www.electronicbookreview.com (accessed April 2002).

————. 2002. *Cognitive Fictions.* Minneapolis: University of Minnesota Press.

Taylor-Guthrie, Danille, ed. 1994. *Conversations with Toni Morrison.* Jackson: University Press of Mississippi.

Trinh T. Minh-ha. 1982. *Reassemblage.* New York: Women Make Movies..

————. 1985. *Naked Spaces—Living Is Round.* New York: Women Make Movies.

————. 1989. *Woman, Native, Other: Writing Postcoloniality and Feminism.* Bloomington: Indiana University Press.

————. 1991. *When the Moon Waxes Red: Representation, Gender, and Cultural Politics.* New York: Routledge.

———. 1992a. *Framer Framed*. New York: Routledge.

———. 1992b. "Naked Spaces—Living Is Round." In *Framer Framed*, 3–47. New York: Routledge.

———. 1992c. "Reassemblage." In *Framer Framed*, 95–108. New York: Routledge.

Turing, Alan M. 1950. "Computing Machinery and Intelligence." *Mind* 59:434–60. Reprinted in *Computers and Thought*, ed. E. Feigenbaum and J. Feldman. New York: McGraw-Hill, 1963.

Turkle, Sherry. 1995. *Life on the Screen*. New York: Simon and Shuster.

Varela, Francisco J., Evan Thompson, and Eleanor Rosch. 1991. *The Embodied Mind: Cognitive Science and Human Experience*. Cambridge, Mass.: MIT Press.

Virilio, Paul. 1994. *The Vision Machine*. Bloomington: Indiana University Press.

Weiss, Beno. 1993. *Understanding Italo Calvino*. Columbia, S.C.: University of South Carolina Press.

Werthheim, Margaret. 1999. *The Pearly Gates of Cyberspace: A History of Cyberspace from Dante to the Internet*. New York: W. W. Norton.

White, Michele. 2006. *The Body and the Spectator: Theories of Internet Spectatorship*. Cambridge, Mass.: MIT Press.

Wiener, Norbert. 1948. *Cybernetics, or Control and Communication in the Animal and the Machine*. Cambridge, Mass.: MIT Press.

Wolfe, Joanna. 2004. "'Ten Minutes for Seven Letters': Song as Key to Narrative Revision in Toni Morrison's *Beloved*." *Narrative* 12, no. 3: 263–79.

Wood, Paul, Francis Frascina, Jonathan Harris, and Charles Harrison. 1993. *Modernism in Dispute*. New Haven, Conn.: Yale University Press.

Woodmanse, Martha. 1994. *The Author, Art, and the Market: Rereading the History of Aesthetics*. New York: Columbia University Press.

Woolf, Virginia. 1976. *Mrs. Dalloway*. London: Grafton.

Young, Robert. 1990. *White Mythologies: Writing History and the West*. London: Routledge.

INDEX

Danielewski, Mark, 73
de Certeau, Michel, 34
Deleuze, Gilles: *Cinema I*, 45–46; *Cinema II*, 49; and the concept of difference and repetition, 4–5; and Felix Guattari, 37, 70, 104
Dirlik, Arif, 133n1
Douglas, J. Yellowlees, 135n13

ecological consciousness, 129, 130
Eisenstein, Sergei, 44–45

Feenberg, Andrew, 119–20
Fisher, Caitlin, 20
Foucault, Michel, 135n16

gender, 1, 2, 39, 55, 112, 136n5.1; construction of, 6–7; in Jackson's *Patchwork Girl*, 64; in Malloy's *its name was Penelope*, 60–61
Gilbert, Sandra, 64
Guattari, Felix. *See* Deleuze, Gilles: and Felix Guattari
Gubar, Susan, 64
Guyer, Carolyn, 19; and Martha Petry, 36

Handley, William, 134n11
Haraway, Donna, 116, 136n5.1
Harpold, Terence, 31, 36
Harries, Karsten, 135n4.1
Harris, Paul, 51–52
Harvey, David, 133n6
Hayles, N. Katherine, 23, 37, 133n4; and constructions of the post-human, 114–16; and flickering signifiers, 35; and media-specific analysis, 22, 76, 91; and multi-course literature, 75

Holmes, Janet. *See* Strickland, Stephanie: and Janet Holmes
Hutcheon, Linda, 16, 62, 104

India, 119, 128, 129, 130
Irigaray, Luce, 6

Jackson, Shelley, 2, 7, 19, 57, 58; and L. Frank Baum's *Patchwork Girl of Oz*, 65, 66; and *Patchwork Girl*, 63–71; Shelley's *Frankenstein*, 63–66, 67–68, 71
Jameson, Fredric, 32, 33
Joyce, James, 11
Joyce, Michael, 19, 27, 31, 36; *afternoon, a story*, 19–20

Kittler, Friedrich, 35
Koolish, Lynda, 17, 134n9
Kuhn, Annette, 62

Landow, George, 36, 103
Lawson, Cynthia. *See* Strickland, Stephanie: and Cynthia Lawson
Lefebvre, Henri, 56
Lyotard, Jean François, 12

Malloy, Judy, 2, 7, 19, 57; *its name was Penelope*, 58–63; and Shelley Jackson, 57–58, 71
Malmgren, Carl, 133n7
Manovich, Lev, 73
McHale, Brian, 133n5, 135n15
media-specific analysis, 22, 76, 91
Memmott, Talan, 20
Minsky, Marvin, 113
modernism, 15; and fictional space, 133n7; and modernist writers, 11
montage, 17, 103; cinematographic,

"The Ballad of Sand and Harry Soot," 76–91; and Janet Holmes, 76, 77; and Cynthia Lawson, 23, 76, 91, 92, 95; *V: Losing L'una/ WaveSon.nets*, 91–101

technocratic narrative, 2, 8, 26; and counternarratives, 130; and narratives, 27; and technocratic logic, 127, 128, 129
technotext, 22, 23. *See also* cybertext
third space, 8, 36; in Bhabha, 42–43, 46

topological space, 2, 33, 35, 39, 94; in de Certeau, 34; of experience, 3, 9; perspective of, 13; in Trinh's films, 44
Trinh T. Minh-ha, 2, 9, 40, 59; and black screens, 48–49; and fragmentation, 32–33; and hypertext narrative, 30, 32; and identity, 39; and montage, 43–44; and negative space, 46; *Reassemblage* and *Naked Spaces*, 46–55

Woolf, Virginia, 11

(continued from page ii)

JAISHREE K. ODIN is professor of interdisciplinary studies at the University of Hawai'i.